T0320552

Advances in Underwater Technology, Ocean Science and Offshore Engineering

Volume 10

Modular Subsea Production Systems

ADVANCES IN UNDERWATER TECHNOLOGY, OCEAN SCIENCE AND OFFSHORE ENGINEERING

CONFERENCE PROGRAMME PLANNING COMMITTEE

Ian Bell, *Subsea Developments Ltd* (Chairman)
Peter Christie, O.S.O., *Department of Energy*
Dr Chris Hedley, *BP International Ltd*
Jean Pritchard, *Society for Underwater Technology* (Conference Organiser)
Dr Bill Supple, *J. P. Kenny and Partners*

Advances in Underwater Technology, Ocean Science and Offshore Engineering

Volume 10

Modular Subsea Production Systems

Proceedings of an international conference (The Modularisation of Subsea Production Systems for Deep Water Application) organized by the Society for Underwater Technology and held in London, UK, 25–26 November 1986

Published by
Graham & Trotman
A Member of the Kluwer Academic Publishers Group

First published in 1987 by

Graham & Trotman Limited
Sterling House
66 Wilton Road
London SW1V 1DE
UK

Graham & Trotman Inc.
101 Philip Drive
Assinippi Park
Norwell, MA 02061
USA

ISBN 086010 832 5 (Vol. 10)
ISBN 086010 922 4 (Series)

British Library and Library of Congress
Cataloging in Publication Data is available.

Typeset by Bookworm Typesetting, Manchester

Printed in Great Britain by the Alden Press, Oxford

Contents

PART 1

PART II

PART III

PART I

1

BP's Diverless Subsea Production System Project (DISPS)

B. S. Green,
Manager, DISPS Project, BP Petroleum
Development Limited

It is BP Exploration's declared objective to develop a diverless subsea production system (DISPS), to bring on stream a subsea field in water depths beyond diver access by the mid-1990s. With application centred on UKCS 9th Round licenses, the DISPS Project team within BPX has started on a programme of engineering, development and testing to provide the necessary technology. In the early stages of Phase 1 of the programme, equipment reliability has been emphasised with work on component qualification and upgrading, reliability modelling and system testing.

The system of modules has been configured for maximum reliability and independent retrieval of least reliable components. Optimization of the system will continue throughout the Phase 1 programme. The system design includes several features intended to aid operation of the remote production facility, such as downhole pressure monitoring. Concentric tubing hangers are involved to ease the problems of orientation and guidance of drilling and workover risers in deep water. Various contracts are in place with designers of equipment, handling systems and control architecture, and further contracts are in preparation for component development.

Advances in Underwater Technology, Ocean Science and Offshore Engineering, Volume 10: Modular Subsea Production Systems
© Society for Underwater Technology (Graham & Trotman, 1987)

Plans for Phase 1 continue after component development with onshore and offshore testing of modules and handling systems, to be completed in 1989. In Phase 2, a prototype system will be installed and used to produce oil in diver-accessible depths, probably exporting to a fixed platform. This full-scale trial will give BP operational experience over several years, and the confidence to specify a diverless system for production to a fixed or floating structure by 1995. Work is continuing to review DISPS application to a range of conditions and to ensure that the work programme correctly reflects BP exploration and development strategy.

EXPLORATION OBJECTIVES

It is BPX's declared objective to develop a diverless subsea production system for operational service in the mid 1990s in water depths in excess of 350 m. In the region west of Shetland, BP holds licenses under the UK 9th Round to drill as operator and in other blocks as co-venturer. Many of these blocks lie on the continental slope where depths between 300 and 800 m are typical. Clearly, our ability to exploit any hydrocarbons found in these technically difficult regions relies on having the production technology in place. One of the immediate consequences of the deepwater exploration programme is the step into diverless technology at the c. 350 m limit; at least, the economics of diver access at c. 400 m become very unattractive.

In view of the cost of the deepwater diverless technology the field sizes in prospect must be large and flow rates high to justify the long development programmes. Looking back to 1965 at over 300 wildcat wells drilled by the industry in depths ranging from 200 to over 2000 m, about 60 have been in water depths exceeding 600 m. Very little success has been achieved in depths exceeding 600 m. Consequently, BP's subsea production system technology is centred on 400 m water depth.

In addition to the UK west-of-Shetland prospects, there are possible applications on the continental slopes of northern Norway, the west of Ireland and in similar continental slope locations worldwide. The DISPS development programme, spanning approximately 10 years, is intended to provide BP with the confidence to pursue a deepwater oilfield development in the mid- to late-1990s. This chapter describes the DISPS concept, which is based on modular principles, and how the preferred configuration of modules has been built up to provide the basis for an extensive development and testing programme. The chapter goes on to discuss the design philosophies and premises, and how the project is controlled to draw all the inputs together successfully. Some of the optimization aspects are discussed including the wider R&D framework that BP is pursuing in deepwater technology. The concluding remarks draw attention to the continual need to review BP's exploration objectives to ensure that the technology pursued under DISPS remains in line with reservoir and field requirements.

DISPS CONCEPT DESCRIPTION

DISPS is a Diverless Subsea Production System being developed by BP for the exploitation of offshore hydrocarbon accumulations in water depths greater than 350 m. Design is centred on 400 m with a further conceptual understanding of system requirements in 750 m. Figure 1 illustrates the proposed modular system of components configured in an 8-slot template.

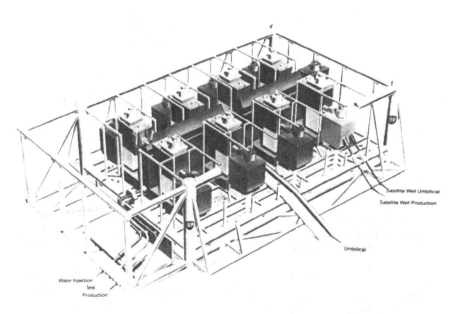

Fig. 1   Eight-slot production template

The system under development consists of a multiwell modular template. The design basis provides for interconnected modules, which can be retrieved to the surface for repair or replacement of components. Figure 2 shows how the Xmas tree, process control, flow control and isolation valve modules have been configured to permit easiest retrieval of least reliable modules. It is envisaged that well intervention for various reasons will require retrieval of the Xmas tree more frequently than the other modules, and so the tree is arranged to be independently retrievable and it is connected to the other modules by a horizontal connector.

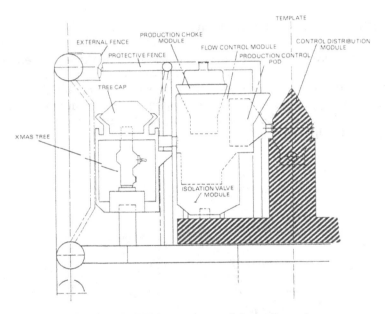

Fig. 2   DISPS base case module configuration

A single control pod is arranged to be retrievable independently with the process control (i.e. choke) module. It is connected to the choke module by a vertical connector and to the control manifold by a horizontal connector. Satellite well tie-in modules can take the place of wells, and these remote wells will be provided with means of connection for control umbilicals at the template. Various combinations of template and satellite wells will be possible within DISPS. In the limiting case, an all-satellite-wells configuration would imply a manifolding function for the template with a configuration similar to Fig. 3. Satellite well designs, methods of control and hook-up of flowlines and control umbilicals are being developed under DISPS' scope of work. The alternative case of individual flowlines carrying well fluids back to the processing facility and with chokes at surface will also be examined and designs prepared.

The DISPS system is to be suitable for oil and gas production with or without water injection, subject to field requirements. The implications of gas lift and chemical injection will be evaluated for impact on the design of modules, Xmas trees, manifolds and flowlines. It is envisaged that most combinations of all these facilities will be required in eventual DISPS applications.

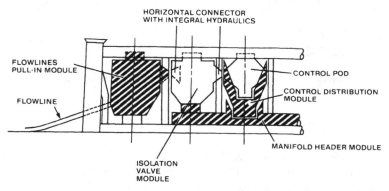

Fig. 3   DISPS concept collector manifold

The DISPS base case design assumes control of individual wells by chokes with pressure and position monitoring on the template and with commingled export of well streams back to a processing platform facility perhaps 10 km distant. Equally important, export to a floating production facility such as the BP SWOPS vessel is under evaluation (Fig. 4). In each of the two cases, flowline lay and tie-back techniques are part of the DISPS design brief. It is intended that simultaneous drilling and production will be possible on the same template, and the DISPS system design includes suitable features to provide for this. Similarly, workover and production will be possible. Design of the well facility for DISPS begins at the bottom of the well with a pressure sentry, continues through the completion design, includes a concentric tubing hanger and will specify the interfaces concerned with normal production or workover conditions. Pigging systems are being evaluated and designed both for frequent and infrequent use and also for the flowline installation phase.

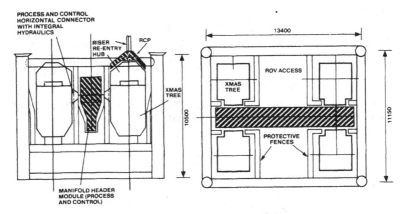

Fig. 4   DISPS concept riser base (SWOPS concept)

THE ORIGINS OF THE PROJECT

Since 1984, a major R&D effort has been applied to subsea production systems, using both external and internal resources. In May 1985 a contract was awarded to FUEL to develop initial designs for two scenarios – a simple 4-slot and a more complex 8-slot Diverless Template/Manifold (DITEMA). In parallel with this, but working independently, a small BP taskforce was set up to review subsea completion techniques. In the event, the study scope was taken further than originally planned and an alternative template concept was produced. This was titled "A New Approach to Subsea Production Systems" (NASPS).

Results of both studies, each supported by physical models, became available during September 1985 and it was immediately clear that each concept had significant advantages in some areas. Accordingly, supplementary work was started to optimize the various inputs and to attempt to integrate these into a single concept, which was to become DISPS. DISPS would be progressively worked up into a prototype design, and ultimately into a truly diverless field development project. By 1 January 1986, the supplementary work had advanced to a point where the basic design philosophy was starting to emerge, although many detailed points remained to be resolved. FUEL were recommissioned to address these unresolved points, taking up many of the NASPS features.

The NASPS work largely stemmed from the Dyce Operations Centre and naturally embodied a lot of operational expertise and experience. During the DISPS concept development a concerted programme of review by Dyce Operations Staff has occurred. Many of the basic operating philosophies with a large impact on template system design, including shutdown, isolation, maintenance and inspection, have been assembled in full consultation with the Operations Centre. Figure 5 shows how the DISPS concept has been developed and defined with the assistance of Operations Staff and the Engineering Technical Centre Staff since the strategy was set out in 1984.

The diagram shows how the strategy gave rise to specifications and workscopes for earlier studies by FUEL, the in-house study, and their joint reappraisal under the FUEL supplemental contract. The current status is shown with most of the major design contracts in place and the Operability Group providing guidance to designers.

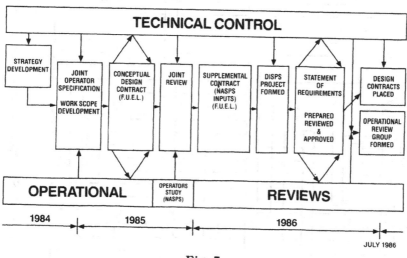

Fig. 5

PROJECT OBJECTIVES

In summary, the objectives of the DISPS project are to develop confidence, by 1995, in the diverless subsea production system principles, equipment and techniques that BP would use to exploit an oil reservoir in 400 m of water, with an understanding of the 350-750 m water depth range.

The project is divided into two phases, Phase 1 (shown in Fig. 6) being broadly concerned with the major engineering design, develop-

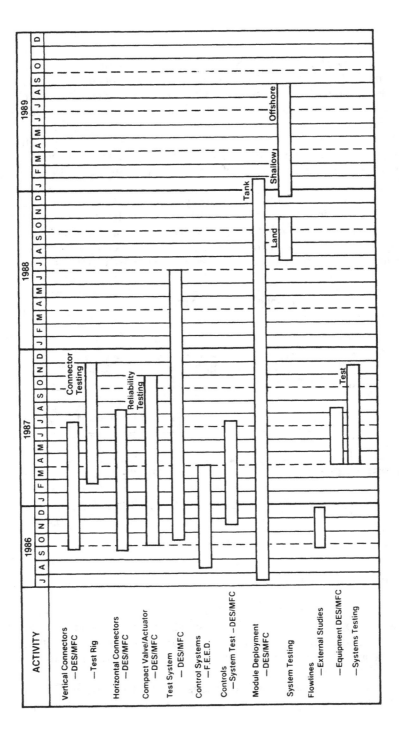

Fig. 6 DISPS Phase 1 work programme (revival August 1986)

ment and testing of components and systems, to be completed by 1989. Phase 2 concerns testing and evaluation of an oil-producing prototype to gain experience of operating the system without diver assistance, with all the normal risks, problems and strictures of oil production in an offshore environment, to be completed by 1995.

Phase 1 is first concerned with confirming the proposed range of components and techniques requiring development. Some components will be available while others will require either upgrading or development. Preliminary designs for these components will then be prepared. These will be followed by detailed designs and manufacturing of components in successive stages punctuated by design reviews, and contract awards to most suitable suppliers.

The Phase 1 objectives are primarily concerned with achieving the most reliable components suitable for DISPS specifications through qualification and testing. Each component development programme starts with qualification of available equipment involving reliability testing in some cases at BP's Sunbury Research Centre.

Desirable and undesirable features will be identified and a new specification will be drawn up taking account of these. Where believed necessary, new components will be constructed to the DISPS specification and tested further. In some cases the new component design and specifications may be shelved until Phase 2 calls for the component to be procured for use in the prototype trial.

Later in Phase 1, system testing is planned which is intended to assess the techniques of module connection, guidance systems and control facilities. The system testing will call up components fully engineered only where design details affect system test results. Thus the system test may require well developed modules and certainly multiported horizontal and vertical connectors, but the Xmas tree could be a dummy block without valves. The systems test specification is FUEL's main brief and relies on inputs from all other contractors. YARD and GEC are concerned with module guidance, and control systems respectively. There will also be a large reliance on the connector designer contractors, as designs emerge. The system tests will be conducted on land initially, followed by increasingly deeper, and environmentally more difficult, water conditions. Since many ROV tasks are being specified for DISPS, affecting inspection, override actuation and valve operations, the system tests will include a considerable programme to prove suitable ROV tooling.

In parallel with the component qualification and testing programmes, a reliability modelling assessment is being prepared. Module configuration will be tested to check that the design corresponds to the optimum use of modules given their relative reliabilities. The model will be extended in successive phases to introduce the concept of availability of the DISPS production system and the effect of intervention vessels for workover, maintenance and inspection.

Phase 1 will have either provided directly the designs and systems necessary for Phase 2 or will have identified existing technology suitable for DISPS purposes. In Phase 2, the Project Team will have identified a suitable test site with the best infrastructure to carry out a fully operational prototype trial. The programme assumes identification of a BP-operated site and will require an Annexe B submission. The trial will last for several years to gain exposure to full weather, planned maintenance and inspection cycles. The selected site will be at diver-accessible depth, but it is hoped that the Phase 1 specified equipment will be fully operable according to the DISPS diverless objectives.

The system will encompass all equipment, including flowlines from the base of a fixed platform or riser base to the well. Well servicing will be included in the Phase 2 trial scope of work. Work is under way to identify suitable test sites and possibly to offset the cost of the Phase 2 trials with oil produced by the DISPS production system.

DESIGN PREMISES AND PHILOSOPHIES

The DISPS system is being designed according to a set of design premises which carry over from earlier work, and are supplemented by an increased emphasis on quality assurance and operability. In addition to the obvious non-intervention by divers, is the premise that all equipment will be wet and exposed to hyperbaric pressure. This gives rise to many implications, not least the requirements for electrical connectors. Inductive connectors will clearly dominate in this condition, but conductive connector technology is rapidly advancing towards an acceptable proven state for deepwater applications and is under consideration for some purposes. The depth range from 400 m to 750 m also affects some mechanical equipment. The power ratings for actuators will vary with ambient pressure and, while designs are centred on 400 m, the implications for deeper water will be investigated.

Probably the most significant design premise concerns the decision to undertake all equipment maintenance at the surface by retrieval of the modules. It is intended that all non-structural modules will be retrievable to the surface for repair, with varying degrees of difficulty, most unreliable modules being configured for most simple removal and exchange.

The service-life target for DISPS will be 20 years wherever possible. Some components will not last for 20 years and a maintenance and inspection philosophy is being developed to ensure satisfactory performance. Naturally, intervention policy, the type of vessel needed and its frequency of use are all factors to be assessed in the cost of intervention versus specifying improved component reliability.

Out of the maintenance and inspection philosophies will emerge certain design constraints and direction to ensure proper access by ROV-conveyed tools or inspection equipment. In some cases it may be appropriate to install inspection monitoring equipment if this can be made more reliable and cost-effective than intervention type inspection techniques. It is DISPS' declared intention to minimize template size consistent with access for maintenance and inspection. In some cases it may be possible to eliminate the need for heavy lift vessels.

In order to ensure that all avoidable equipment failures are eliminated, a high emphasis is being placed on quality assurance. In addition to inspection engineering being applied to designs for improved inspectability, a similar effort will be applied to the highest possible quality of materials, workmanship and cleanliness. Suppliers and manufacturers will be subject to stringent quality assurance plans and quality control procedures under BP supervision. Quality plans are being developed in conjunction with detailed component designs and for procedures forming part of the system tests.

The DISPS Operability Review Team is continuously assessing operability issues down to procedural level to assist design and development. The group is guided by and assists in developing suitable philosophies for normal or upset and test conditions. Central to these philosophies is a loss control philosophy. This takes account of isolation policy, leak test policy, flushing policy and control philosophy. The number and types of testable barriers between well fluids and the environment are the cornerstone of the loss control philosophy. The unique circumstances of a DISPS system require a clear understanding of possible problems so that facilities can be designed and constructed to assist an operator to safely and effectively control production from a remote installation. Where equipment failure requires vessel intervention the system must shut down in a controlled and ordered way. The control system architecture will therefore take full account of the loss control philosophy.

Development of any of the key philosophies and policies is a demanding and onerous task involving collation of many viewpoints. However, attention to the most detailed procedural analysis greatly improves design specifications and directions given to design contractors.

No less a task is the input to drilling equipment interfaces where drilling or workover will take place. It is a basic design premise that drilling intervention will take place in conjunction with continued production. All drilling technology input under DISPS resides in the project or can be called upon in the Dyce Operations Centre. Contractors need to be carefully briefed on all drilling equipment interfaces. A large body of loss control policies peculiar to DISPS drilling operations and workover equipment has to be developed. The technology ranges from guidance equipment and provisions for landing of 200 tonne BOP's through to control and test facilities on Xmas tree caps. All of the downhole completion design suitable for production and injection modes falls within the project Drilling group's brief to develop policy and philosophy for DISPS.

DESIGN AND DEVELOPMENT METHODOLOGY

The strategy prepared for DISPS development in 1984 had divided the workscope into segments suitable for available contractors to undertake at successive conceptual and detailed design stages. The Phase 1 work programme in Fig. 6 shows how the dominance of front-end design and conceptual design extended well into 1987. Approval by the DISPS project between stages is ensured before further commitment and to ensure that due account has been taken of all the inputs. All the present activities feed into the template systems test, where the span of land and progressively more severe wet testing extends from mid-1988 to late 1989.

The main design contractor, FUEL, is responsible for template system design and also coordinates all other design inputs affecting template design. FUEL also assembles all documentation and maintains records. FUEL's primary design task is to identify the scope and extent of a template system test. It is intended that the test specification will include dry and wet testing, and offshore UKCS testing will be fully representative of the DISPS system's ultimate duty. The test relates primarily to accuracy of fit, successful operation of guidance systems for modules, and control of module retrieval, testing and deployment systems.

FUEL's work thus relies heavily on work by the other major

contractors, YARD and GEC, and the component suppliers. YARD is developing guidance systems and vessel specifications using nominal module sizes and weights. YARD's work is conditioned by certain first-pass design constraints such as use of vessels of opportunity, and using proven components and systems with least engineering risk. YARD has therefore been concerned with identification of concepts, analysis of forces, assessing the capabilities of typical support vessels and looking towards the techniques of umbilical handling in deep water. The conceptual work will gradually give way to more detailed analyses of vessel motions and capabilities of heave compensation systems, all of which will influence FUEL's work on guidance and capture of modules at the template.

GEC has been awarded the Control System design contract: they will develop the Control System architecture and examine issues such as the selection of connectors where inductive versus conductive connector choices may arise or where, for example, the effectiveness of many inductive connectors in series suffer unacceptable attentuation.

The control system will be multiplexed electrohydraulic without further hydraulic back-up and so reliability assessments of systems and components feature prominently in the design process. GEC will examine whether components testing and upgrading of control components should be undertaken in common with the methods of component upgrading over the rest of the Project. Included in the Control System evaluation is acoustic telemetry for satellite wells, and BP is also party to the S.W.A.I.-GEC joint industry programme to investigate acoustic methods. In support of hydraulics design, GEC has subcontracted Dowty Boulton Paul.

In addition to the major systems designs, the component development programmes have been or will be progressed with suitable suppliers and designers. It is intended that a typical component development programme would follow the sequence set out below.

(1) Survey industry and other operators' reported experiences to qualify broadly suitable components.
(2) Review designs of preferred selection of components.
(3) Identify design improvements by reliability comparisons and testing under hyperbaric conditions.
(4) Specify improved designs for DISPS applications.
(5) Specify quality plan.
(6) Design and build new component.
(7) Test for improved reliability.
(8) Confirm designs for DISPS application.

The extent of the programme adopted in the case of each component will vary depending on the Projects' judgement of the suitability of the component for DISPS and confidence in its reliability. Components such as connectors, for example, do not exist for DISPS purposes and so stages (1) and (2) are more concerned with contractor qualification for design and construction than with quality surveys of components that are currently used industry-wide.

The testing stage of component selection will involve comparative performance trials over many accelerated cycles on the Sunbury test rigs. This sort of testing lends itself particularly well to valve, choke and actuator testing. The larger components such as module connectors, pods and flowline connectors will initially undergo factory acceptance tests, and will then be subject to very much more exhaustive performance tests, in some cases with the manufacturers and in others at Sunbury.

Flowline connectors and flowline lay techniques are being pursued under a separate contract. This programme follows the sequence of flowline lay method investigations, equipment design and specification, manufacture and testing. The design work is complemented by Sunbury scale-model tests and will lead into full scale trials offshore later in the Phase 1 work programme.

The Phase 1 programme relies heavily therefore on component development achieving the published plan. Any delays will immediately begin to affect progress of the template test programme. Undoubtedly, a large part of the Projects' resources are devoted to maintaining adequate progress and to providing effective liaison between contractors.

DESIGN OPTIMIZATION

The present DISPS configuration derives from the foregoing studies which set out to optimize the relative content of modules and their positions. The DISPS project largely adopted the then preferred configurations as the vehicle for further development under the DISPS Phase 1 and 2 plans. Thus, the present configuration relies on largely conceptual studies and has not yet taken full account of the now-scheduled reliability work which may confirm the existing DISPS scheme or possibly modify some aspects of it. The design has been and will continue to be a play-off between the arguments of simplicity with minimal failure modes, and enhanced operability which typically calls for more isolation, more access and more information. The impact of any changes arising from either side of the argument will be highly disruptive to the progress of the DISPS system design and so very strict review procedures have been imposed.

The work to confirm equipment and module configuration is the subject of a reliability model being developed on BP's behalf by Bradford University. Various module configurations will be studied using available equipment reliabilities. Also, a target reliability will be derived, given a target mean time to failure for the DISPS system. This work is being complemented by a Hazard and Operability study conducted in-house.

In common with all BP projects, a six-stage safety review programme is applied. Teams of specialists review designs and make recommendations to project managers: their brief is essentially to achieve safety and good operating practice. The process has begun in DISPS to examine the provisional process and instrumentation designs under the formal hazop procedure. The report by the hazop team will be used to amend designs where appropriate in the interests of safe practice and to perhaps increase the objectivity of the design of the DISPS system.

One of the modules which has already received a lot of attention concerning its optimum location has been the control pod. Figure 7 shows how several possible locations would have resulted in a different balance between numbers of electrical and hydraulic connector groups. The accessibility of the pod also varies such that it can be retrieved independently or combined with other modules. Until GEC reports on the control architecture design the pod will remain on the Process Control (choke) module where it is connected vertically, and horizontally, to the control distribution module (see Fig. 8). The GEC study will be taking account of the more detailed control requirement of other modules, notably the Xmas tree and satellite trees, to examine whether a more reliable but equally effective configuration exists.

In order to balance out the arguments of improved operability tending towards increased complexity, the DISPS Project is examining how ROV technology can complement installed facilities. Valve actuation, for example, if required very infrequently, lends itself to ROV control rather than to extra hydraulic pathways through already crowded module connectors.

ROV technology is being examined, starting from a list of probable tasks. The extent to which these will be adopted will be a judgement between the competing technologies involved and their state of development. It can be seen from the following provisional task list how ROV technology would need to be extended to cope with both greatly increased weights and complexity, compared with its present

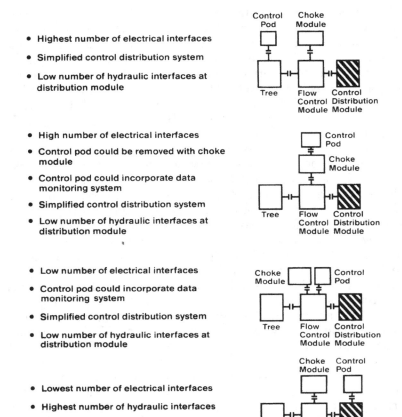

- Highest number of electrical interfaces
- Simplified control distribution system
- Low number of hydraulic interfaces at distribution module

- High number of electrical interfaces
- Control pod could be removed with choke module
- Control pod could incorporate data monitoring system
- Simplified control distribution system
- Low number of hydraulic interfaces at distribution module

- Low number of electrical interfaces
- Control pod could incorporate data monitoring system
- Simplified control distribution system
- Low number of hydraulic interfaces at distribution module

- Lowest number of electrical interfaces
- Highest number of hydraulic interfaces
- Separate power and signal connections required for the data monitoring system

Fig. 7   Control distribution configuration – optimization of pod location

status if, for example, the lightweight cap installation and removal task shown below was extended to full-pressure retaining caps.

Provisional List of ROV Tasks

- observation (and recording) of module landing sequence
- actuation of connectors via hot stabs

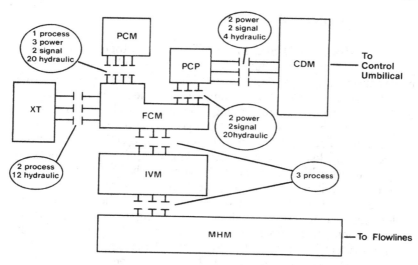

Fig. 8   DISPS base case control pathways

- connector integrity testing
- valve overrides – mechanical
- lightweight cap installation or removal
- seal replacement
- replacement of subsea electronics modules on Xmas trees
- isolation valve module actuation
- jetting and cleaning
- inspection tool deployment
- acoustic transponder changeout
- connector lockdown
- temporary repair of leaks

During the system tests the contribution by ROV technology will be a key aspect of the in-water evaluation programme.

The whole of Phase 1 is concerned with the optimization of component and system design. The judgements primarily concern fitness for purpose. As the optimization process moves forward, cost will play an increasing role, and perhaps experience with DISPS "Mk I" may be an acceptable compromise with which to go forward into Phase 2. Further experience of hydrocarbon production operations in Phase 2 is likely to advance the optimization process significantly towards DISPS "Mk II".

PROJECT CONTROL

In March 1986, the DISPS Project Team came together as a small group geared to establishing the workscope and controlling work programmes and packages let to either contractors or in-house design groups. The team comprised engineers responsible for engineering coordination, drilling technology and for applications engineering. The size of the team, together with support from BP's Engineering Technical Centre, served the Project's needs for the period of workscope definition, culminating in a Statement of Requirements (SOR) document. In view of the previous activities aimed at describing and defining the DISPS system, this document became more than just a set of technical requirements and design philosophies. It was very much closer to being a system description and, in some areas, included specifications for components. The DISPS Project Team took the SOR, developed workscopes accordingly and started to put further studies in place and award contracts.

By *no* means are all the philosophies and policies needed for DISPS finalised. DISPS is a development project and the technical solutions to its goals will not be established until testing is complete and to the Project Team's satisfaction. Inevitably, changes of direction will occur and earlier supposed solutions to problems will be dismissed. Control of the project therefore is not simply concerned with advancing through a closely specified scope of work. The workscope will and must change as the understanding develops. The Project Team has more recently increased in numbers to manage the project with a greater technical control, following contract awards. More technical judgement is being called for within the project to achieve timely review of work packages.

The size of the Operational group in the DISPS team has also been increased to improve turnaround of reviews. Following the setting up of an Operational review group, the dissemination of its findings and recommendations to the design teams has placed a particularly heavy workload on the Project Team. The Operational review group drawn from Dyce Operations Centre, BPX's Technical Directorate in London, and the DISPS Project has been called upon to develop the DISPS design scope down to procedural level. By drawing on direct experience of subsea operations in Magnus, Buchan and from engineering support to maintenance and inspection programmes generally, it is intended that design proposals be tested for their operability before those designs are approved by the Project. The process is laborious and demanding but is seen by the DISPS Project as a key input to the design process.

In order to keep the various design contracts in step, turnaround of reviews has become critical. The Engineering Services Coordinator is supported by planning and cost-control services to ensure that the Project's progress is monitored and controlled to meet its targets.

Although FUEL, the major design contractor, is briefed to coordinate technical inputs from all other contractors and therefore to pull together the system design, DISPS project approval according to the approved SOR is essential. Approval also takes account of advice from discipline specialists in BP's Engineering Technical Centre, and where design issues affect drilling technology or operations all advice is in-house derived.

The drilling technology group within the DISPS team is responsible for well completion design and resulting implications for Xmas trees and for defining module interfaces. Between the technical discipline engineering support from the Engineering Technical Centre and the Operability Review Group and Drilling Department, the DISPS Project can call on the support of an effective and balanced team.

SUPPORTING R&D

In the earlier stages of the DISPS concept development from 1984 to early 1986, all the R&D topics now incorporated in DISPS Project scope were part of a broader subsea R&D strategy. The activities now falling outside DISPS continue, with work in progress at the BP Research Centre, Sunbury, or with sections of industry in joint venture activities. Noteworthy amongst these has been the two-phase pump, being developed in partnership with Stothert and Pitt and other operators. A subsea-specified version of the two-phase pump could eventually find an application as part of a future DISPS system.

In selecting the R&D topics which now fall within DISPS scope of work, many decisions have had to be made where the alternatives were equally valid for inclusion as DISPS base case topics. Notable among these is the type of flowline system now being developed in detail. The single flowline concept has been selected while the bundled concept has been set aside. However, the implications of a bundled system will be studied conceptually. This method of proceeding has been similarly applied to the issue of wellhead and Xmas tree working pressure rating. 10 000 psi wellhead and 5000 psi trees have been specified for DISPS while the higher-rated 15 000 psi and 10 000 psi cases respectively will be studied conceptually.

Some of the items set aside in the DISPS base case selection feature in the BPPD Norway R&D portfolio. The DISPS project maintains a regular dialogue with Norway to ensure that, where a topic could readily provide technological support to a future or alternative application of DISPS, then suitable guidance is given to the R&D control officers in Norway.

In the case of the flowline option previously discussed, BP is pursuing an extensive design and testing programme on bundled flowlines, with Kvaerner. Pigging systems are being designed by Seanor with a 12-inch flow system planned for construction and extensive testing.

Other topics include intervention methods of choke and pressure sensor changeout, and conductive connector studies in conjunction with work at the BP Research Centre, Sunbury. In order to support future DISPS applications to a range of deepwater prospects, further technologies have been identified for interfacing with DISPS. These include, for example:

- support from a TLP (Fig. 9)
- support from a tanker system (Fig. 10)
- risers, buoys and swivels
- DP systems

Topics such as these form part of the ongoing BP R&D portfolio in the area of production technology.

APPLICATIONS FOR DISPS

The DISPS Project continues to review prospects and production scenarios to give insight and more definition to the DISPS programme. If a clear prospect existed with oil in place, fluid quality, number of wells and the nature of adjacent support facilities, then the DISPS task would be clear, and other options and technologies would be set aside for the moment. Our present understanding dictates that we examine many avenues and options for DISPS exploitation with equal attention.

Figures 9, 10 and 11 illustrate how various configurations of DISPS could be assembled. For example, where lift capability from wells is limited it will be necessary to continue individual flowlines back to the processing facility and choke flow at surface. The design of the module configuration would consequently be simplified but a new

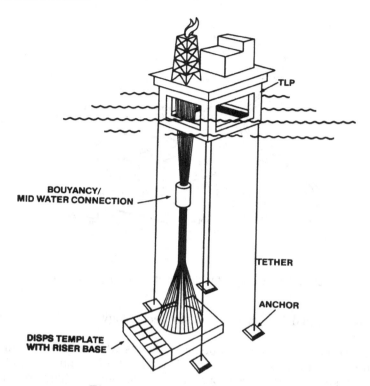

Fig. 9   DISPS development to TLP

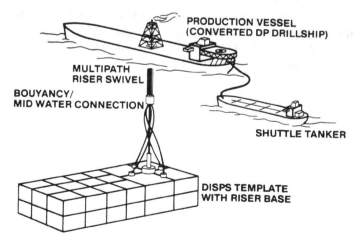

Fig. 10   DISPS development to floating production vessel

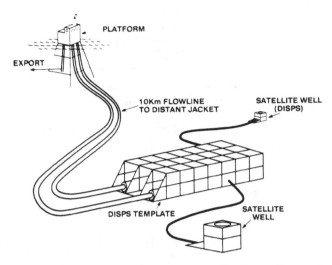

Fig. 11   DISPS development in deep water, flowing to distant platform

emphasis would be applied to riser technology, including buoyant systems.

A further avenue which is being studied is whether DISPS can be made economic in shallow water. Shark infestation and tidal factors render some areas hazardous for divers. All of these possible applications are under continuous review to ensure that the DISPS programme adopts the correct emphasis within the overall exploration strategy. It has to be remembered that DISPS is an R&D project and must continuously reflect the overall BP exploration and development strategy. The application studies are seen as a key input to the strategy review process.

2

Campos Province Deepwater Exploitation Perspectives

S. Armando,
Early Production Systems Group, Production
Department, Petrobras, Brasil
and
A. R. Cvijak, Subsea Engineering Group, R&D
Center (CENPES), Petrobras, Brasil

INTRODUCTION

Thirteen Years Made Short (1973-86)

Until shortly before the 1973 oil price increase, offshore activity was in its earliest stages: at that time only large fields were considered to be economically feasible for exploitation. In 1979, large platforms made their debut in the North Sea in the Ekofisk, Frigg, Brent, Piper, Forties, Ninian, Statfjord and Beril fields and, in the USA, in the Cognac and Hondo fields.

To cope with the tight economics of small reservoir discoveries, the Early Production System (EPS)/Floating Production Systems (FPS) concept arose independently in Argyll (1975) in the North Sea, Enchova (1977) in Brasil, Castellon (1979) in the Spanish Mediterranean, and later in various other fields such as the Garoupa-Namorado one atmosphere EPS, and the Pampo, Dourado, Casablanca, Cadlao, Buchan, Bicudo, Badejo and Tazerka wet completions. The 1970s were doubtlessly the era when reality caught up with imagination and when exploration, drilling and exploitation techniques made a gigantic technological step moving to deeper and deeper offshore fields.

Advances in Underwater Technology, Ocean Science and Offshore Engineering, Volume 10: Modular Subsea Production Systems
© Society for Underwater Technology (Graham & Trotman, 1987)

Deep Water Means . . .

Before and during the 1970s, anything beyond 100 m W.D. was included under the heading "deep waters". The evolution of saturated diving techniques as well as remote operated vehicles (ROVs), electronic remote control systems, connectors, etc., made the 100 m figure change to the present approximately 300 m diving limit.

There are already diving companies experimenting with different gas mixes and testing as deep as 400-500 m, as well as off-the-shelf commercially available ROVs that operate at depths in excess of 1000 m. In Brasil, Petrobras presently considers "deep waters" to start at 300 m, though it is expected that saturated diving depth limits could reach 420 m with no significant technological changes.

SUBSEA INSTALLATIONS

Floating Production Systems

Though pushed by the oil price increase, the offshore oil industry move to deeper waters put at constant risk the economic feasibility of (up to then) well proven engineering solutions. Exploring, drilling and exploiting techniques used for 100 m water depths were simply not usable (as they were) at 500 m: technical and economical barriers proved to be hard to overcome.

One of the results of the above-mentioned feasibility red light was the startling development of not only more complex and at the same time more reliable subsea equipment, but also new ideas and concepts regarding offshore and subsea transportation, installation, operating and maintenance techniques and procedures. In the Southern Hemisphere Petrobras is actively participating in this breakthrough. A good example of this evolution is the Early Production System (EPS): Petrobras has been one of the very first oil companies to test this concept and intend to make intensive use of it in the development of deepwater fields.

The first Petrobras subsea completion was approved by the board of directors in 1975: it was the Garoupa-Namorado one atmosphere EPS, with chambers installed in W.D. varying in the range of 120-160 m, plus a dry-manifold and two tanker-mooring towers (one for a converted processing ship and the other for the load-out one), and making extensive use of multiplex electronics. However, due to various problems, by the time this complex system began to operate in 1979, the Enchova wet subsea completion had been installed and had already started operation back in 1977. At that time, the only other "deepwater" EPS in operation was in the Argyll field in the North Sea. Today (November 1986), Petrobras have:

- 11 FPS in operation;
- 2 FPS being installed;
- 7 FPS being designed.

Wet Xmas-trees

The first Petrobras wet Xmas-tree was installed during April 1979 over well 1-RJS-38, in the Campos Basin Bonito field (Fig. 1). A further 76 wet Xmas-tree and 7 dry one-atmosphere chambers have been installed since (Table I).

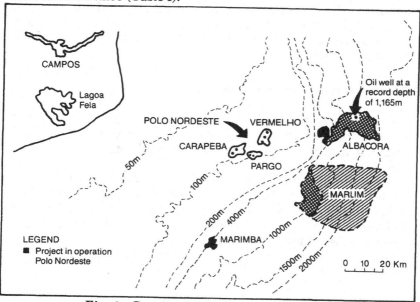

Fig. 1   Developed fields in the Campos Basin

TABLE I.
Wet X-trees.

Depth	Installed	To be installed	Being studied
0– 49m	3	6	–
50– 99m	8	4	–
100–199m	46	34	10
200–299m	18	15	7
300–399m	2	9	–
+400m	–	7	3

Note: Marlim and Albacora are not being considered.

The EPS effort in the Campos Basin led to a series of subsea completions setting new water-depth world records, as listed below:

Well	W.D.(m)	Date
1-RJS-38	189	April 1979
3-BO-3-RJS	208	October 1982
7-CO-4-RJS	226	July 1983
1-RJS-54	254	August 1983
4-RJS-232	293	December 1983
3-PU-2-RJ	307	September 1984

The present world wet subsea completion water-depth record is held by 1-RJS-284, standing at 383 m W.D. in the Marimbá field, and also in Campos Basin. Before the year's end, the same field's 1-RJS-294 should be completed at 413 m W.D. Both Xmas-trees are of the diverless/lay-away type.

Manifolds and Template-manifolds

The idea behind the use of subsea manifolds is to reduce the number of risers from the seafloor up. Of course, the price paid for this load reduction on the floater is the cost of the additional subsea valves, chokes, piping, etc.

The goal of the template-manifold is to eliminate costly subsea connections between satellite Xmas-trees and manifolds, therefore saving not only miles of flexipipe but also the time and money involved in laying and connecting piping. However, the additional expense is that of directional drilling.

An update of Petrobras manifolds and template-manifolds is as follows:

Field	Type	Quantity	Number of slots	Status
Bonito	m. template	1	10	operating since 1979
Linguado	manifold	1	10	operating since 1982
Garoupinha	manifold	1	6	operating since 1986
Viola	manifold	1	8	installed/start-up in December 1986
Bicuco	manifold	2	8	to be installed in March 1987
Albacora	manifold	2	6	under procurement
Albacora	m. template	3	21	being designed
others	manifold	4	8	under construction
others	manifold	8	8	under procurement
others	manifold	8	8	being designed

It is worth noting that all installed units are directly hydraulically controlled, but deepwater ones should be electronically multiplexed.

Pipelines And Risers

Most Petrobras EPS use flexible pipes, with a flowline bundle formed by:

- 1 × 4 in production
- 1 × 2½ in annulus
- 1 × electrohydraulic flat-pack

Flexible pipes are commonly used to diameters up to 12 in. Rigid steel pipelines have been only used (a) when diameters exceed 10 in and/or (b) in projects where exploitation is expected to last for more than 10 years.

It is a general belief that flex-lines have the following main advantages:

- short time to install
- easily removable
- easily re-installed
- no problems if the sea-bottom is not flat.

An extensive use of flex-pipes has been made in the Campos Basin EPS developments, including their use as production risers. An update of flexible pipe utilization follows:

- approximately 600 km in operation
- the purchase forecast (over 3 years) of 40-60 km/year
- re-installation of 8-10 km/year

DEEPWATER FIELDS

Considering 330 m to be the present threshold for the deepwater concept, the presently known Petrobras deepwater fields are Marimbá, Albacora and Marlin.

The Marimbá Field

Marimbá is 15 km north-east of the Enchova Central Platform, at water depths varying between 350 and 550 m. The field's estimated reserves are around 10×10^6 m^3. The oil layer is in Cretaceous sandstone at 2750 m. At present, only the 1-RJS-284 wet Xmas-tree completion record-setter is in production in Marimbá.

The field will be exploited through an FPS integrated to the already existing Piraúna FPS (Fig. 2). The design allows production through up to eight satellite oil wells plus three water injection wells. The estimated flow will total about 40 000 bopd.

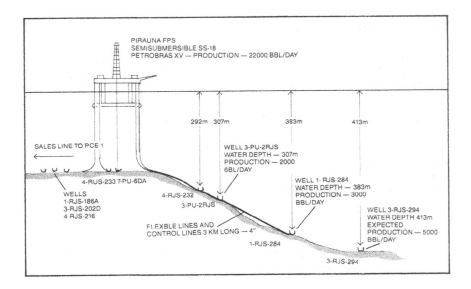

Fig. 2   Piraúna floating production system

The Albacora Field

Albacora is 48 km north-east of Garoupa, in water depths varying from 250 to 2000 m. Geologically, it has been only partially defined, but it is known to have seven pay-out sandstone layers/reservoirs between 2600 and 3200 m. Estimated reserves exceed 150×10^6 m^3 enclosed in a 150 km^2 area.

In order to adapt to the water depth characteristics of this field, exploitation will occur in three phases, using step-by-step technological up-grading criteria.

The first production phase consists of 6 to 12 subsea wet Xmas-tree completed satellite wells, all connected to a single manifold through a flexible bundle.

Oil/water/gas separation will be done at a converted 55 000 t.d.w. tanker, yoke-moored to a CALM-type swivel buoy. Oil export is to be to the existing Garoupa fixed steel Central Platform. Gas will be flared. One of the most important goals of Phase 1 is to collect well production data for more accurate reservoir knowledge.

Phase 2, in due course, intends to produce from the Carapebus sandstone layer. About 60 wells are to be drilled using three to six template-manifolds, to be connected to the manifold already mentioned in the Phase 1 description. The same floating facilities will be used as in Phase 1.

Phase 3 intends production of the whole field by means of well clusters similar to those used previously. Plans include installation of a fixed steel platform for oil processing and gas compression. It is estimated that about 150 wells will be producing by that time, the total flow being around 250 000 bopd.

TABLE 2
Albacora field development plan

Phase	Number of wells	Q_0 (bopd)	Water depth (m)
1	6–12	15 000	250–450
2	50–60	60 000	250–450
3	about 150	200 000–300 000	350–2000

The Marlin Field

Marlin is at 35 km east of Garoupa Central Platform and has been found to cover an area of about 150 km^2, ranging from 600 to 1200 m W.D. It's reserves are estimated to be 250×10^6 m^3. At present, the field is under intense geological study: It's development is planned to start in 1988.

PLANNING FOR THE FUTURE

The Critical Path

Many solutions have been proposed recently for each of the problems related to the offshore oil business. Each one has it's own particularities.

However, to a great extent, some of the solutions are still considered to follow the deepwater techno-economical critical path:

- floating units and their mooring/DP systems
- risers
- pipeline installation, connection and repair
- subsea equipment operation and maintenance
- safety and reliability of all the above items

Original Ideas Welcome

Even though only very recent projects involve the production of offshore fields located at water depths in excess of 400 m, it is believed that existing technology can reach the 600 m limit with no significant technological breakthroughs.

However, it must be stressed that field exploitation economics can be severely at risk if the use of existing concepts and technologies is to be considered.

In fact, Petrobras projections show that design, construction, transportation, installation plus operating costs can be three to four times heavier when considering the exploitation of "modern concept" deepwater fields with fixed structure technology, when compared to the cost of the same activities at the previous deepwater threshold.

We therefore believe that, from both technical and economical points of view, the trend in the near future for deepwater offshore developments will be the thorough use of floating units equipped with processing or pre-processing facilities, exporting to complete processing plants installed either on fixed offshore structures or on shore.

A second but more technologically distant alternative is to install the processing facilities directly on the sea bottom. We believe the present pace of oil energy consumption will oblige the oil industry to move even deeper, until a time technological breakthroughs will make it safe and economically feasible to install the above-mentioned equipment on the sea bottom, whether wet or dry.

Finally, we believe that in order to keep the present standards of human and operational safety at a premium and, at the same time, to minimize capital risks and, therefore, transportation, installation, operating and maintenance costs, emphasis must be given to the research, development and engineering of original system and equipment conceptions.

PART II

3

Modularization of Trees

A. Fletcher and I. Pirie,
Vetco Offshore Limited

INTRODUCTION

Subsea production trees have now been used for the control of oil and gas wells drilled under water for some 20 years. The first trees were simple affairs, designed for use in shallow water, being installed and possibly operated by divers. These trees were derived from existing surface equipment using stacked valves with the minimum of integration.

As oil discoveries have moved to deeper water and harsher environments on marginal sized developments, there are greater reasons to consider subsea completed wells as an alternative to fixed platform developments. Two other major considerations influencing the balance in this direction are the use of subsea multiwell manifolds and floating production facilities. Multiwell subsea manifolds can be used to produce either to adjacent fixed platforms or floating production facilities.

Advances in Underwater Technology, Ocean Science and Offshore Engineering, Volume 10: Modular Subsea Production Systems
© Society for Underwater Technology (Graham & Trotman, 1987)

DEFINITION OF THE MODULAR CONCEPT

There are two basic concepts that describe the modular tree system. The first (Fig. 1) is the construction concept where a manufacturer uses a number of proven design modules to tailor a subsea tree design to suit a customer's particular needs.

The second is the operational requirement to split the tree into individual remotely retrievable modules, so that certain sections of the subsea tree can be retrieved to the surface for replacement or repair without disturbing the other sections of the system. Neither of these concepts are mutually exclusive.

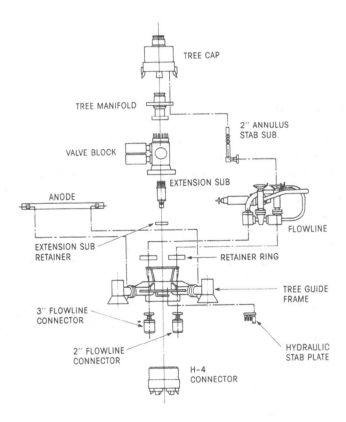

Fig. 1   Tree unitization

Tree Function and Interface

The function of a subsea tree can vary depending on the nature of the system to which it is matched.

In its most simple form, the tree is only required to divert flow from the well at the wellhead and to isolate the well in the event of a failure downstream of the tree. In this mode the tree will have a simple control system and will not be operated very frequently; thus it may almost be considered a passive component.

As experience and confidence in subsea systems grow, subsea trees are being required to perform functions in addition to the simple system tests. Trees can be fitted with subsea chokes to actively regulate flow and valve system control, and the injection of wax- or scale-inhibiting chemicals either downhole or into the flowline.

As part of a subsea production system, the tree is no longer a passive component, but may be the only means of isolating one well from another prior to co-mingling at the template for export. Along with this increase in function and status within the system will come an increase in operating cycles and risk of failure.

Interface

The trees' primary interface is with the wellhead, which not only supports the BOP (during drilling) and casing strings, tubing hanger and tree, but is the ultimate pressure-containing barrier between the well and the environment.

In the North Sea, most wells were initially drilled with 21¼ in × 2000 MWP and 13⅝ in × 10 000 psi dual stack system. However, laterally most wells are drilled with 18¾ in systems in either 10 000 psi or 15 000 psi ratings. Some wells are drilled with 16¾ in × 10 000 psi systems, normally from ship shaped dynamically positioned vessels or to suit floating production systems where the 16¾ in systems offers the advantage of lower stack and riser weights.

The tubing hanger is designed to run through the BOP with the downhole assembly, land, lock and seal in the wellhead where it provides a connection between the tree, the production tubing, subsurface safety valve controls, downhole pressure and temperature sensors and the tubing/oil string annulus. An interface between the tubing hanger and the tree is provided by stabs mounted on the base of the valve block. These are made up simultaneously with the landing of the tree on the wellhead, and should provide metal-to-metal sealing with the seal pockets or receptacle on the tree and tubing hanger.

CONSTRUCTION MODULES

A subsea tree is basically an assembly of seven sub-assemblies:

- wellhead connector
- flowline connector
- flowline pipework
- valve block
- tree mandrel
- tree cap
- tree guideframe

Each of these sub-assemblies can be chosen from a range of options and integrated to produce the required tree design; of these seven sub-assemblies the wellhead connector, tree mandrel and tree cap are well defined components which do not change in design from one project to another.

Wellhead Connector

The wellhead connector is designed to attach and seal the tree to the high-pressure wellhead housing, using its integral hydraulic system to energize the locking dogs on the external profile of the wellhead. The ring gasket is retained inside the connector and effects the seal between the tree and the wellhead. The connector will also have secondary hydraulic releases, mechanical releases and provision to externally test the wellhead gasket.

Flowline Connector

The flowline connector system has a tremendous range of designs and complexity, from the simple diver made-up system using swivel flanges, to the remote diverless flowline pull-in and connector system. The major considerations in choosing a flowline connector system are water depth (whether it is practical and economic to use divers), and the type of completion, whether it is template multiwell system, or satellite trees.

A simple flanged flowline make-up can be used on either satellite or template trees where diver operation is acceptable. To make the tasks easier for the diver, provision should be made to relieve him of any heavy physical task by providing attachment points for lifting and pulling equipment, and hydraulic pull-in and flange make-up tools.

A vertical hydraulic flowline connector will allow diverless flowline connection to pre-installed pipework on the guide base or template. This requires that the pre-installed pipework is terminated in an accurately positioned mandrel with which the flowline connector will mate when the tree is landed on the wellhead. A limited degree of freedom in both connector and mandrel is required to allow self-alignment.

The most sophisticated flowline connection is the one that requires diverless flowline pull-in and, because of this, there must be strong justification for considering the use of such a system, i.e. operating beyond diver depth with an adequate number of completions over which to amortize the cost of the tools.

Flowline Pipework

The design of the flowline pipework is not a contentious matter and is simply a matter of routing the various lines in a conventional and practical manner. There are two major influences on the flowline pipework:

- whether it is requirement to incorporate as many valves as possible in the valve block to minimize the number of connections, or
- the number of auxiliary functions such as chemical injection and crossover lines

Valve Block

In order to maximize the intrinsic safety it is the trend, for the North Sea, to include as many as possible of the tree valves in a single block forging. This is done at added cost and risk during the manufacturing operation. Valve blocks vary from the simple block with three production and annulus valves to a block with 10 valves all incorporated in a single block forging.

Tree Mandrel

The tree mandrel is provided to attach the tree running tool and tree cap and is a well defined assembly. Associated with the tree cap is the control transfer from production to workover mode. This is achieved by control cartridges placed round the mandrel through which each tree function is piped. The stabs on the tree running tool isolate the production control system and allow workover control through the tree running tool, while the stabs in the tree cap complete the control loop through the cartridges allowing control of the tree from the platform.

Tree Cap

The tree cap acts as a protection cover for the tree mandrel and a secondary pressure-containing barrier above the swab valves. The caps also forms part of the control transfer system as detailed in the previous section, by carrying the control transfer stabs.

This is a well defined section of the tree, with the main options being whether it is hydraulically operated or manually operated by the diver.

Tree Guide Frame

The tree guide frame, as the name implies, is provided to guide the tree while it is run, or retrieved from, the well and in its simplest form of four arms with guide cones this is all it does. However, in practice, it may take on a more complex shape and role by incorporating the following functions:

- protection of the tree during transport and running or in operation
- an attachment point for ROV operation of valves
- support for a flowline pull-in mechanism
- structural support and guidance for a control package

OPERATIONAL MODULES

In anticipation of the requirement to produce oil from fields in water depths beyond those projected for practical diver intervention, there has been work carried out to design systems for remote intervention. The solutions considered for this include:

(a) Manned intervention with encapsulated one-atmosphere systems having the option of wet or dry environments, where the operators transfer from mini submarines to large chambers containing all the equipment.
(b) Manned intervention using a one-atmosphere suit such as "Jim" or "Wasp", carrying out work on the equipment in the marine environment.
(c) Unmanned intervention, using ROVs, carried out in the marine environment.

(d) Unmanned remote-operated tools carrying out specific tasks on dedicated sections of the system.

(e) A modular system in which all sections of the system incorporating active components are packaged as modules which may be retrieved to the surface for replacement and repair.

Option (a) is expensive because the whole system (consider a template of eight trees) has to be incorporated in a chamber to withstand an external pressure of 1400 psi for water depths of around 3000 ft.

Option (b) requires that large "walkways" be provided for access and the number of tasks that can be completed is limited.

Both (a) and (b) present hazards to the operations resulting in stressful operating conditions. Option (b) will also have a very limited operating time on bottom.

Options (c) and (d) suffer similar disadvantages to (b) in that the space required for access and the tasks that can be carried out are limited.

The modular systems allow the equipment to be arranged with the minimum of spacing and, as all the active components can be retrieved to the surface, there is very little limitation to the scope of work that can be carried out.

The basis of the modular approach is that all the active components are contained in modules and a failure may occur in a section of the system; thus this section may be removed without compromising the other modules.

To carry this philosophy to its logical conclusion, all active components in a subsea system should be modularized and, considering a template system to be the optimum for a deepwater production system, all the template manifold valves should be integrated into manageable modules, an example of which is shown in Fig. 2. Prime candidates for failure are: sensors, chokes, flow control/isolation valves and, in particular, wing valves. Consideration should be given as to whether it is desirable to pull a module, to replace a particular component or whether the system can safely operate with the failed component. Normally, the system would tolerate failure of some of the sensors without serious risk to safety or the operation of the system, whereas failure of the wing valve or choke on a tree does present a serious risk. Therefore, to make the system manageable, the critical, high-risk components (such as choke and wing valve) should be integrated in one module which is easily retrieved with the minimum disturbance to other modules.

Most modular systems to date are based on using the standard

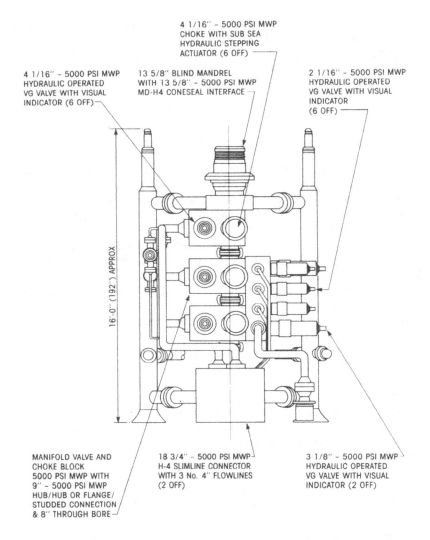

4 1/16" - 5000 PSI MWP
CHOKE WITH SUB SEA
HYDRAULIC STEPPING
ACTUATOR (6 OFF)

4 1/16" - 5000 PSI MWP
HYDRAULIC OPERATED
VG VALVE WITH VISUAL
INDICATOR (6 OFF)

13 5/8" BLIND MANDREL
WITH 13 5/8" - 5000 PSI MWP
MD-H4 CONESEAL INTERFACE

2 1/16" - 5000 PSI MWP
HYDRAULIC OPERATED
VG VALVE WITH VISUAL
INDICATOR
(6 OFF)

16'-0" (192") APPROX

MANIFOLD VALVE AND
CHOKE BLOCK
5000 PSI MWP WITH
9" - 5000 PSI MWP
HUB/HUB OR FLANGE/
STUDDED CONNECTION
& 8" THROUGH BORE

18 3/4" - 5000 PSI MWP
H-4 SLIMLINE CONNECTOR
WITH 3 No. 4" FLOWLINES
(2 OFF)

3 1/8" - 5000 PSI MWP
HYDRAULIC OPERATED
VG VALVE WITH VISUAL
INDICATOR (2 OFF)

Fig. 2   Template manifold module for six wells

API guidewire centres for use with drillpipe to recover the
equipment (e.g. Shell's UMC & Elf's Skuld and East Frigg project).
This leaves scope for development of a guidelineless system possibly
using cable to run and retrieve the modules. This approach is
required for deepwater systems (>73 000 ft), where guidelines are
impractical and a cable running system is required to give reasonable
running and retrieving times.

Christmas Tree

The standard subsea Christmas tree can be considered as a module because it can be recovered by disconnecting it from the wellhead and flowline connector after having set wireline plugs in the tubing hanger. For deepwater conditions this would hardly be acceptable, as it would involve two round trips with workover riser and two wireline operations. If it were possible to retrieve the faulty wing valve or choke in a module run and retrieve on cable there would be no need for wireline intervention, as there would be adequate well protection using the master valves left with the lower section of the tree on the well.

The tree can be modularized in two ways. One method is that the tree is stacked vertically (see Fig. 3) with the choke module and wing valve on the top section, or alternatively it can be split with the two sections side by side. The first option gives a rather tall tree, but simplifies control and alignment problems, while the second option gives a shorter system which could also incorporate the flowline isolation valves.

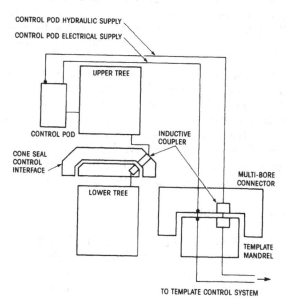

Fig. 3.

One of the key components not normally found on a standard subsea completion will be multibore connectors with 5 in and 2 in production and annulus bores which will also be capable of making up multiple hydraulic and electrical connectors.

Control Module

The Satellite Control Module (SCM) is a fully integrated electrohydraulic control module, designed to control all subsea valve functions associated with the satellite well, and to monitor the status of these valve and various well condition sensors. There is one SCM fitted to each well. This will be mounted on a base which will be fitted to the tree frame so that it can be plumbed to the tree control without the need for additional connectors.

The SCM contains single atmospheric electronic package for the control of solenoid-operated valves, and for the receipt and transmittal of commands and data to and from the control centre. Remote sensors are connected to the SCM electronics via inductive couplers through the base of the SCM.

The SCM is separately retreivable from the rest of the subsea system. All hydraulic connections to the hydraulic umbilical and to the function lines take place through the bottom part of the module. Self-sealing poppet connectors with primary metal-to-metal seals and back-up "O" rings are used to connect the module hydraulic lines to its mounting base. The electrical signal and power connections are made through the base of the module.

All SCM components are contained in a pressure-compensated housing, the hydraulic sub-assembly being located in the lower part of the module and the electronics in the upper part. Through the control umbilical, electrical power and signals between the platform and the SCM are transmitted by an electric multicore umbilical.

The design of the SCM allows diverless deployment. Therefore, the SCM may be run, located and clamped remotely using an appropriate running tool and multifunction connector.

Accumulator Assembly

An integral part of the subsea control hydraulic supply system are two subsea accumulator assemblies. The assemblies employ bladder type accumulators to store high-pressure fluid to the tree. Each assembly is mounted on the tree near the control module mounting base.

Pressure Sensor

The housing assembly for the remote pressure sensor is constructed from 316 S12 stainless steel and is designed to withstand the external water pressure.

Production Manifold Module

The subsea manifold will be a retrievable assembly of valves with their associated controls and pipework, assembled in a suitable package to allow the assembly to be run and retrieved using standard systems available aboard drilling vessels. Figure 2 shows a design of one such package which contains the chokes, isolation valves and injection system for six trees designed to be run on the standard 6 ft radius, API guide post. This design uses a multiblock design and was derived after examining the two possible extremes of an assembly of discrete components and a system where all the valves and chokes were incorporated in one block.

The advantages of this method of packing allow the equipment to be handled from standard drilling vessels with only vertical entry and connections to the subsea template. This allows a compact template design, thus minimizing installation problems associated with large offshore structures.

CONCLUSION

Two definitions of the modular concept have been examined in this discussion; the constructional modular concept and the operational modular concept. Both have relevance in the concepts for deepwater subsea production systems. The constructional modular system gives ease of design by allowing the configuration of field-proven modules to suit conventional tree or production systems destined for deeper water.

Modular trees are perceived as a requirement for deepwater production systems and they do offer answers to the intervention requirements, but in parallel with this, a great deal of work is being done with ROVs to produce tooling packages to carry out subsea intervention work.

The final and most desirable solution to the problem is to improve the system reliability to a level at which intervention is not a prime requirement.

4

Mechanical Connectors

A. Gledhill and B. Hart,
Cameron Iron Works

INTRODUCTION

Mechanical connectors are present in many different areas of subsea oilfield equipment ranging from drilling BOP stacks through Christmas trees and flowlines to the latest concepts in modular completions.

They perform a very important interface and absolute integrity is required to prevent pollution. This is particularly so with their extensive use in modular template completions when many more such connectors are used than would be the case with more traditional methods.

PURELY DIVER-INSTALLED CONNECTOR

In the case where diver involvement is acceptable, a very cost-effective and reliable type of connector is that of the type shown in Fig. 1. This uses individual tie-down screws which force

Advances in Underwater Technology, Ocean Science and Offshore Engineering, Volume 10: Modular Subsea Production Systems
© Society for Underwater Technology (Graham & Trotman, 1987)

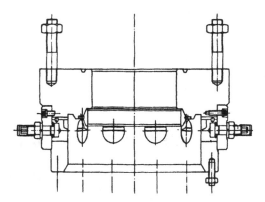

Fig. 1   Mechanical wellhead connector

wedge-shaped segments radially inwards against the tapered shoulder of the wellhead housing.

Its main advantages are low initial cost and reliability with no hydraulic components to denature after prolonged installation. However, it is labour intensive and obviously not suited to use in deep water and requires longer to make up tight than hydraulically actuated types.

WELLHEAD HYDRAULIC CONNECTORS

A type of connector using hydraulically actuated collet fingers is shown in Fig. 2. In this design, a number of hydraulic cylinders drive a tapered sleeve which acts against the collet fingers, driving them onto the mating hubs.

This design is very versatile and is equally suited for use as a drilling connector or a Christmas tree connector.

In the unlocked position the clamp fingers open out at the bottom to provide a funnel which improves guidance when used as a wellhead connector. This feature is particularly useful during guidelineless drilling.

When greater initial alignment is required as in the case of a tree connector, where internal stab mandrels are used, a lower skirt ring is employed to provide initial lateral clearance to prevent damage.

As the connector mechanism relies on sliding tapered sleeves to derive its mechanical advantage, the effects of friction have an important bearing on overall efficiency.

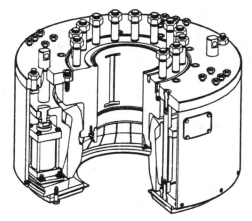

Fig. 2   Model 70 collet connector

For the connector shown in Fig. 1, the mechanical advantage is given as

$$M = \frac{1}{2}\left[\frac{1}{6\tan(a+f_1)\tan(b+f_2)} - 1\right]$$

where a = angle of outer tapered sleeve
$\quad\quad\;\; b$ = angle of wellhead housing and connector body
$\quad\quad\;\; f_1$ = friction angle for sliding sleeve/collet finger interface
$\quad\quad\;\; f_2$ = friction angle for collet finger/wellhead housing interface.

A typical value of a is 4° and of b is 25°. Consequently, if friction is ignored, the mechanical advantage is 14.83:1.

However, the effects of friction greatly reduce this figure such that a practically achievable mechanical advantage is of the order of 3:1. This figure is obtained using an angle of friction of about 8° for both f_1 and f_2, which is equivalent to a friction coefficient of 0.14.

Since friction has such a dramatic effect on efficiency, particular attention is given to the use of low-friction coatings and special lubricants. After a considerable amount of use or after a long period subsea it is unlikely that a connector will perform as efficiently as when new.

Given the practical limitations of the current design an improved way of increasing connector preload is to increase the hydraulic force on the tapered sliding sleeve as on the annular piston hydraulic connector shown in Fig. 3. The annular piston gives a larger surface area than the total for the individual cylinders of the non-integral design of a similar size and operating pressure. Mechanically, this connector is similar to that shown in Fig. 2, but it is more suitable for use when high cyclic bending loads are encountered such as when drilling in very deep water or during wellhead tieback operations. Increased preload prevents joint separation under load and reduces the risk of fatigue failure of the clamp fingers and fretting of the metal gasket.

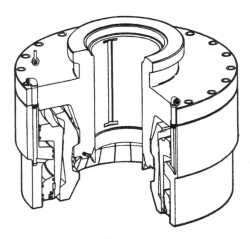

Fig. 3   HC collet connector

These types of connectors are considered to be self-locking, since the angle of the outer tapered sleeve at 4° is less than the assumed friction angle of about 8°.

External mechanical locking devices may be provided, using the standard override rod feature if any vibration is expected which may cause the connector to loosen during installation.

SUBSEA CONNECTORS WITH SEPARATE ACTUATORS

The collet connectors described previously are integral with their hydraulic actuators. This can be a disadvantage where such connectors are immersed subsea for extended periods with the risk of corrosion and damage to the hydraulic components.

A tree connector which works in the same principle as those previously described, although with remote actuators, is shown in

Fig. 4. The hydraulic actuating cylinders are incorporated into the tree running tool shown in Fig. 5. The running tool is shown attached to the tree in Fig. 6 and close-up photographs of the connection between the hydraulic cylinders on the running tool and the extended operating rods of the wellhead connector are shown in Figs 7 and 8.

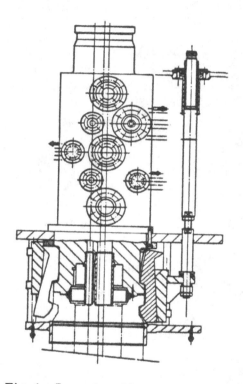

Fig. 4   Connector with remote operators.

A special lockdown mechanism is provided at the top of each operating rod (Fig. 9) to prevent accidental unlocking during service.

After the operating rods have been pushed down by the actuators to lock the connector, on removing the actuation tool, four serrated fingers spring out and engage with a mating fixed collar.

In order to unlock the connector, the actuation tool attaches to the top of the connector operating rod and compresses the serrated fingers, removing them from mesh with the collar. (Fig. 10). The operating rod may then be drawn upwards to unlock the connector.

Fig. 5   Actuation tool   Fig. 6   Assembly of actuation
                                          tool attached

Fig. 7   Locking mechanism   Fig. 8   Locking mechanism

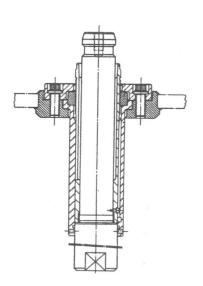

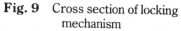

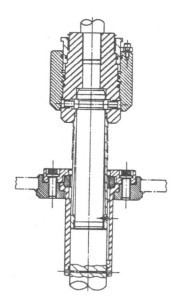

Fig. 9   Cross section of locking
mechanism

Fig. 10   Cross section of
operating rod with actuator
attached

CONNECTORS FOR MODULAR SUBSEA COMPLETIONS

The latest developments in subsea completions employing the modular concept, where subsea production systems are remote from wellheads, require connectors which have multiple bores and are capable of being remotely installed and tested with a high degree of reliability.

A typical small process line connector is shown in Fig. 11. This works on the annular piston concept as described earlier for a wellhead connector, but incorporates two non-integral hydraulic pistons which act as an emergency secondary unlock by means of two rods threaded into the annular piston.

Metal-to-metal seals for each bore are obtained by using ring gaskets squeezed into tapered seats. As the gaskets are flexible in a radial direction when compared with the hubs, they are pressure-energized but they are nevertheless given a substantial initial squeeze on make-up of the connector in order to prevent leakage at low pressure.

Initial alignment on bringing the connector into proximity with its

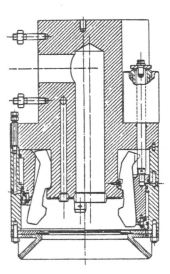

Fig. 11   Annular piston process line connector

mating hub is given by the lower skirt ring which incorporates a guide
funnel. After the funnel has entered over the neck of the hub, two
alignment pins are provided which extend beyond the gaskets and
enter into pockets in the hub to give correct angular orientation
before the gaskets enter their seats.

Most collet connectors were originally intended for use in a
vertical mode and in this situation no special consideration need be
given to circumferential location of the collet fingers. However, in the
horizontal mode, gravity would cause the fingers to bunch together
on the lower side. Although the connector would still function, since
the fingers have a lateral relief provided at the top and bottom,
increased friction arising between each finger would impair the
connector efficiency.

To overcome this problem, a number of small guide blocks are
provided to locate the fingers circumferentially. They are screwed to
the connector body and fit into grooves machined in the clamp finger.

Alternatively, a small grub screw may be provided in the side of
each clamp finger, protruding the same amount as the normal gap to
provide equal finger spacing.

The annular piston is provided with lip seals and bearing rings to
prevent galling. The bore of the cylinder is hard surfaced with either
an electroless nickel finish or a corrosion and wear-resistant weld
cladding.

During installation of the connector and just prior to final make-up, a quantity of corrosion inhibitor may be injected into the clamp finger area from the annulus bore.

A flexible seal ring is provided in the lower skirt ring, to be a snug fit around the hub. This traps the inhibitor and prevents circulation of oxygenated water around the moving parts of the connector.

CONNECTION TOOL FOR REMOTE MAKE-UP OF FLEXIBLE PIPE

Flexible pipe is popular for connecting subsea completions and satellite wells. It is easily installed using remote diverless pull in systems.

To make up a flowline connection to a satellite well, a sheave frame (Fig. 12) is attached to two auxiliary posts mounted on the guide frame. The actual winch is mounted on a rig or workover vessel, the cable passing round the pulley mounted on the frame.

Fig. 12   Flowline pull-in sheave frame

The free end of the flexible pipe has a nose cone to which the cable is attached and is drawn towards and locked into a special retainer which is provided on the well guide base. The sheave frame is then recovered to surface and replaced with the hub make-up tool (Fig. 13) which is also shown during a workshop simulation of installation in Figs 14 and 15.

Fig. 13   Clamp assembly make-up tool

Fig. 14   Clamp assembly make-up   **Fig. 15**   Clamp assembly make-up
tool during workshop simulation   tool during workshop testing

The make-up tool incorporates a device to remove the protective nose cone from the pipeline and also from the inboard hub mounted on the well guide base. The two clamp halves are shown complete with the central plate which carries the seal ring in Fig. 16.

Fig. 16   Clamp halves with seal plate

Figure 17 shows the clamps and seal plate built up into an assembly with the large central screw which actuates and preloads the clamp assembly. The assembly shown in Fig. 17, described from the bottom, is as follows:

(1) The lower clamp half, which is very similar to a standard clamp.

(2) The seal plate, carrying the tapered metal gasket and also providing alignment for both hubs during the make-up process.

(3) The upper clamp half, very similar to the lower clamp half and also carrying the seal plate.

(4) The bridge piece which bears down on the upper clamp half. This carries a large thrust bearing to take the axial load from the central power screw.

(5) The reaction piece to which are fastened the two long studs which pass through the bridge piece and upper and lower clamp halves. The power screw is also threaded into the reaction piece, such that the bridge piece is forced downwards and the studs are pulled in tension. The two rods projecting from the top of the reaction piece are for handling purposes within the make-up tool.

Figure 18 shows the clamp assembly mounted onto the two mating hubs.

The clamp assembly is contained within the make-up tool. After installation of the make-up tool with the protective caps removed from the flexible flowline and the fixed hub, the clamp and seal plate assembly is lowered into place in the gap between the two hubs. The flexible hub is drawn towards the fixed hub by two horizontal hydraulic cylinders, sandwiching the seal plate in the process.

The two clamp halves are then drawn together by means of the single large diameter power screw which is actuated by a geared hydraulic motor.

Once the connection is correctly preloaded, the make-up tool is recovered and reset, ready for the next connection.

Fig. 17   Clamp assembly

Fig.   18 Clamp and hub assembly

SPECIAL FLOWLINE CONNECTORS FOR HARD PIPE

Although flexible pipe conveys certain installation advantages, it has drawbacks during service. It lacks the reliable durability of hard pipe and thus requires more frequent statutory inspection. Once hard pipe is installed, it can be reasonably assumed to have a service life of up to 30 years.

In order to facilitate installation of hard pipe, some means of increasing flexibility is required. This can always be achieved by using either long pipe runs or loops or coils, but the use of pipeline swivels is usually most efficient.

Pipeline swivels require a fine surface finish in order to prevent leakage during service since they are not preloaded. However, they do permit subsequent movement and are thus able to contend with misalignment occurring due to thermal growth or surface movement.

The alignment swivel flange (Fig. 19) allows angular misalignment of up to 10° in any direction and also permits disconnection, albeit with diver involvement but this is acceptable for pipelines laid in shallow waters. The connection uses a standard API ring gasket and when the bolts are correctly torque tightened, the preload in the joint gives a full metal-to-metal seal in the swivel mechanism. However, once the joint is made up, no subsequent movement is possible unless the studs are slackened first.

Where diver involvement is not desirable, an alternative solution is to use the self-aligning collet connector (Fig. 20), which contains all of the beneficial features of the alignment swivel flange combined with remote operation.

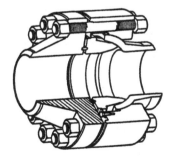

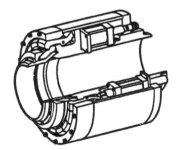

Fig. 19 Alignment swivel flange

Fig. 20 Self aligning collet connector

CONCLUSIONS

Subsea completions are formed of discrete packages of hardware, each dedicated to a particular function, which must be linked together so as to operate as an integral unit.

The interface, whether it be for the passage of production fluids, hydraulics actuation fluids or electrical signals, is provided by the mechanical connector.

The type of connector selected depends largely upon its intended purpose, whether it be made up by divers or operated and installed using remote means. Criteria such as initial cost notwithstanding, the current trend favours fully remote operation as a means of ensuring reliability irrespective of environmental conditions.

Apart from the use of connectors during drilling operations, most other connector installations remain subsea for extended periods. It is essential that they should disconnect reliably when required, such as when equipment is recovered for overhaul. Corrosion in salt water is a serious problem and, although care is taken that the connector mechanical components are adequately protected with special coatings, the use of corrosion inhibitors which displace sea water from the working parts greatly improves reliability. Some damage to protective coatings is usually unavoidable and cathodic protection devices are just as essential overall as are inhibitors.

Special stainless steel welded inlays are usually provided in the metal gasket seats to prevent corrosion. Even a small amount of corrosion occurring at the gasket/seat interface would cause leakage to occur.

5

Performance of Inductively Coupled Electrical Distribution Networks for Power and Signals

R. Phillips and G. R. Clark,
Offshore Projects Group, GEC Avionics

THE REQUIREMENT

Subsea production system hardware has control and instrumentation needs to serve the following functions:

Control of:
tree valves
isolation valves
variable-orifice chokes
pigging/TFL valve diverters
crossover valves

Measurement of:
production pressure and temperature
valve and choke actuator positions
downhole sensing
pig/tool passage detection
hydrocarbon leak detection

The provision of these control and instrumentation needs has evolved over the past ten or so years, towards some fairly standard approaches to implementing the remote control and monitoring function.

Advances in Underwater Technology, Ocean Science and Offshore Engineering, Volume 10: Modular Subsea Production Systems
© *Society for Underwater Technology (Graham & Trotman, 1987)*

Typically, for an electrohydraulic multiplexed control and monitoring system (E-H mux system), distributed control pods (or control modules) provide the location of remotely commanded electronics units which can operate electrical pilot valves and read data from sensors. The electronics units require electrical power and signals to be provided from a surface facility.

ELECTRICAL POWER DISTRIBUTION

Generally, a single seabed umbilical feeds electrical power from a surface electrical power generation and conditioning facility, located on a fixed or floating production facility, often over significant distances, say 5 or 10 or even 15 km (see Fig. 1 for a schematic installation).

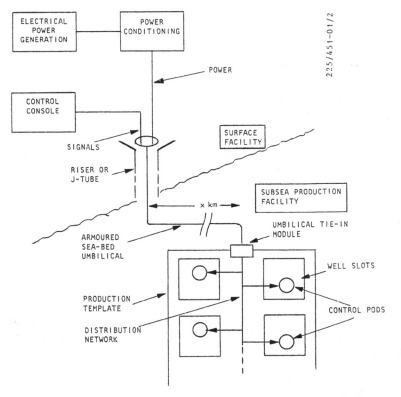

Fig. 1   Schematic installation

The umbilical, which may well carry redundant power feed circuits, "plugs in" to the electrical distribution network on the production template, using some form of umbilical tie-in module/pull-in connector or termination assembly.

The electrical distribution network is installed to feed electrical power (and signals) to distributed control pods which contain electronic multiplex data acquisition and control units. Figure 2 shows a typical pressure-isolating, one-atmosphere enclosure for the electronic circuits, and Fig. 3 shows how this might typically be contained within the envelope of the production control pod.

Fig. 2 Typical pressure isolating enclosure for subsea electronics unit

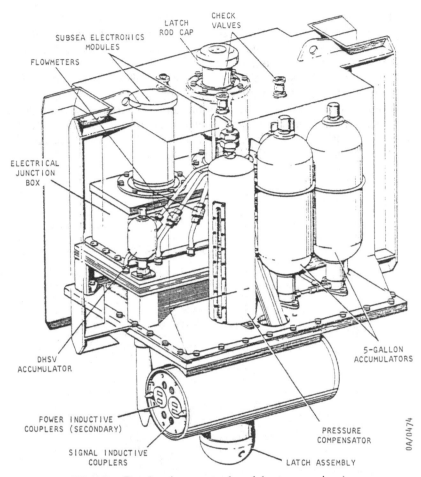

Fig. 3   Production control pod (cutaway view)

    The quantity and location of these control pods is optimized to meet functional modularity requirements, as well as size, weight and installation and retrieval procedures for the control pod. The electrical power drawn at each location on the network is quite small and is determined largely by the number of sensors used for internal monitoring of the pod hydraulic circuits and external sensors (pressure, temperature etc.) on the trees and manifolds, rather than by the electronic unit multiplexing circuits (which are typically implemented using low-power CMOS technology). Various energy-saving strategies are available to reduce the total power drawn by sensors and, although the peak power drawn by solenoid driving circuits for pilot valves can be 10–20 watts, the average power is low since these are typically operated rather infrequently.

The electrical distribution network can typically take two forms, either:

(a) an on-template distribution bus or ring main (see Fig. 4A), or
(b) individual "jumper" connections from the umbilical termination assembly (see Fig. 4B).

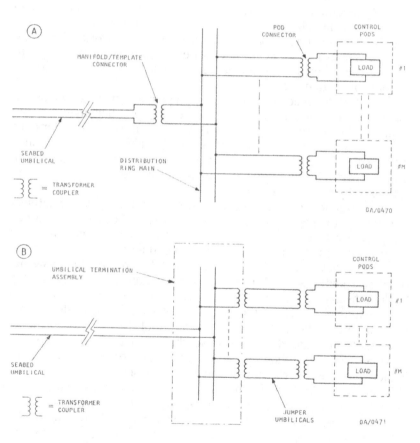

Fig. 4 Alternative distribution networks

Obviously, from an electrical schematic point of view, these configurations are the same except that in case (b) the ring-main or bus is physically made within the umbilical termination assembly.

For system A, assuming first that M is a reasonable number (say 4 to 10), and secondly, design optimization has avoided thermal problems at the large single manifold/template connector, this system is more reliable than B, since there are less transformer couplers than in B.

However, system B has a higher "system reliability" (availability); in other words, B has a higher probability than n out of m circuits are available for use.

Now, the practicalities of remote installation on the sea bed in deep water for a modular production system require several remotely made-up connections as part of the distribution network. The trend towards modular systems for deepwater remote installation tends to more rather than fewer interfaces and each interface requires a remote connection to be made.

For networks which can be made-up using wet/mating conductive connectors the electrical design of the network becomes a volts/amps line losses type calculation. But where inductive couplers are used for connections a more sophisticated analysis is needed, since the coupler behaves as a loosely coupled transformer and is an integral part of the network, the performance of which can be optimized by careful trade-off.

The attraction of inductive couplers is that they are relatively insensitive to gaps and mating misalignments and are hence compatible with the mechanical mating of relatively large chunks of hardware, with relatively loose make-up tolerances, which is the case for modular deepwater remotely installed production systems. Additional features which inductive couplers provide are the facility for live disconnect, since these devices can be made with no metal parts exposed to sea water, and the facility to control disconnect current so that the disturbance on the rest of the network as more, new loads are added (for example during modular expansion) is minimal. (Figure 5 shows typical mating pairs of power and signal inductive couplers for a comparison of relative size.)

There is a view within the industry that networks made up of several levels of series and parallel connections using inductive couplers have, inevitably, a poor performance. This reputation for inefficiency is caused by an imprecise understanding and improper analysis – often caused by extrapolating from the performance of a single unit to multiple units connected in a network. We will show by the two following examples what can be achieved.

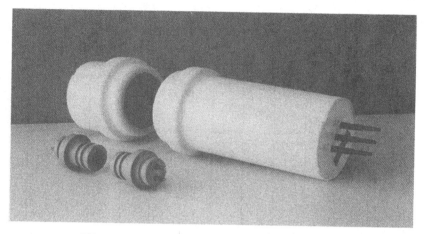

Fig. 5   Power and signal inductive couplers

Figure 6 shows the network as used on a five-slot production manifold. Low power requirements were achieved by having all sensors contained within the control pods. The network's performance characteristics, which were verified on a bench model, are shown in Table 1.

TABLE 1

Parameter	Experimental	Computed
input voltage (VRMS)	309.2	309.2
input current (ARMS)	0.220 lead	0.274 lead
input power (W)	57	55
input power factor	0.838	0.648
manifold input voltage (VRMS)	293.0	298.6
manifold input current (ARMS)	0.662 lag	0.583 lag
manifold input power (W)	50	52.8
manifold input power factor	0.258	0.303
manifold power efficiency (%)	64	69.8

Figure 7 shows the network for a four-slot production template, with power requirements higher than the previous example due to instrumentation external to the control pods. The network's performance characteristics are shown in Table 2.

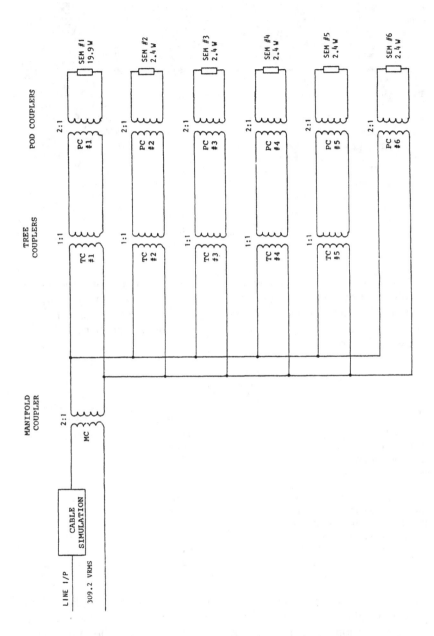

Fig. 6   Power distribution network for five-slot production manifold

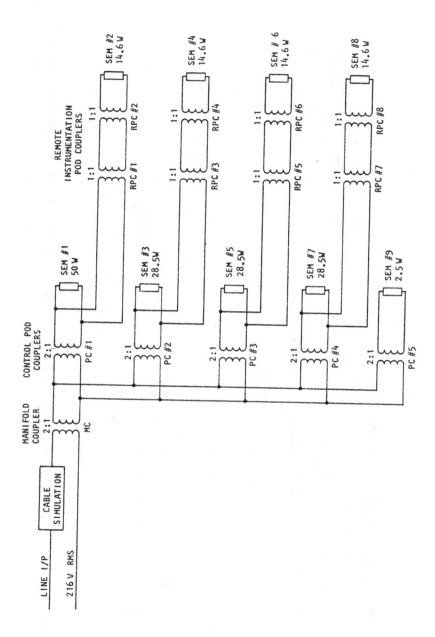

Fig. 7 Power distribution network for four-slot production template

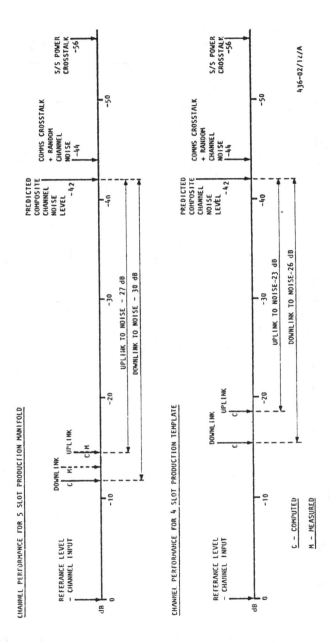

Fig. 8 Communications channel performance – five-slot production manifold and four-slot production template

TABLE 2

Parameter	Computed
input voltage (VRMS)	216
input current (ARMS)	1.602 lead
input power (W)	320.6
input power factor	0.927
manifold input voltage (VRMS)	184
manifold input current (ARMS)	1.434
manifold input power (W)	257.5
manifold input power factor	0.977
manifold power efficiency (%)	76
manifold VA efficiency (%)	89

ELECTRICAL SIGNAL (TELEMETRY) DISTRIBUTION

The equivalent networks for telemetry signal distribution have an identical pattern of series and parallel connections for a particular template geometry.

Through analysis, attention is paid to the attenuation and phase shifts which affect the FSK (Frequency Shift Keying) signals as they pass through the network. Typical attenuation budgets are represented as shown in Fig. 8, these being for the five-slot production manifold and four-slot production template referred to in Figs 6 and 7.

The predicted noise level in the communications channel is plotted at 42 dB down on the reference channel input level. The actual channel attenuation is then plotted as shown to give a signal-to-noise level at the relevant receiver. The two factors, apart from umbilical characteristics, which significantly affect the performance of the communications network are:

- *Coupler axial gap.* A suitably designed mechanical assembly, with couplers individually resiliently mounted, can minimize axial gap which is more critical than misalignment in limiting the number of series coupler connections that can be made.
- *Disconnect philosophy.* It is generally not possible to guarantee a fully populated template in operation because of factors such as early production strategies, routine module maintenance and future expansion requirements. The effect of disconnects on a live network can be minimized by the use of passive distribution networks (PDN). PDNs have an additional feature of providing a degree of fault isolation within the network, and when optimized,

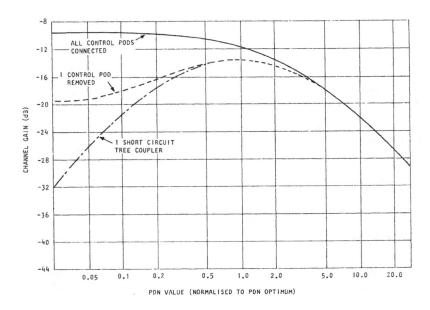

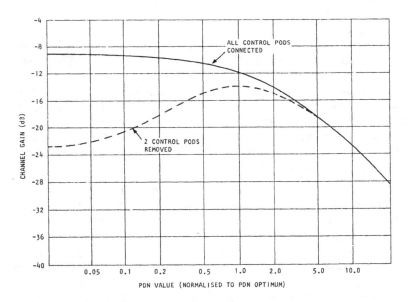

Fig. 9 (*Above*) Downlink channel performance *vs* PDN five-slot
production manifold configuration. (*Below*) Downlink channel
performance – four-slot production template

have a minimal effect on end-to-end network performance. Figure 9 shows the relative affect of the PDN on network end-to-end performance examples for the five-slot manifold and four-slot template.

CONCLUSION

The North Sea can provide examples where inductive couplers have accumulated a very significant amount of device hours and established their track record – the Shell UMC and BP-Magnus fields. These couplers are also qualified for operation at depth (with reference to the Chevron Montanazo programme qualification to 4000 ft). They are mechanically adaptable for diver, ROV and remote make/break situations, and they are cost-effective (certainly in cost-of-ownership terms if not in first cost-capex-terms).

In this chapter we have described some work which demonstrates that, properly optimized, these networks can achieve high performance which is more than adequate for the level of modularization being proposed for deepwater developments by some of the oil industry majors.

PART III

6

Flowline Tie-ins

S. W. Duckworth, W. J. Supple and W. T. Neilson,
J. P. Kenny and Partners Limited

INTRODUCTION

This chapter discusses the variability of tie-in methods and equipment applicable for current and future deepwater installation. The tie-in of a pipeline to a wellhead or template involves many aspects of subsea engineering, ranging from pipeline installation methods through ROV operations to the architecture of the template itself. This chapter attempts to give a reader concerned with template design a better understanding of the problems facing the pipeline engineer in deep water. It also presents current proven limitations for various aspects of diverless tie-ins and provides a checklist for the template designer covering his design elements of the problem.

It is the authors' contention that a "standard tie-in module" is not a practical piece of equipment for the industry to develop because of the variability of the problem: rather, the key common aspects of tie-ins should be identified and equipment/techniques developed and proven that can be assembled to tackle a variety of configurations. This chapter attempts to identify these key aspects to assist designers of future development programmes.

Advances in Underwater Technology, Ocean Science and Offshore Engineering, Volume 10: Modular Subsea Production Systems
© Society for Underwater Technology (Graham & Trotman, 1987)

THE DEFINITION OF DEEP WATER

Deep water is best defined as water beyond the reach of diver intervention. The point at which deep water begins is therefore not fixed; it depends partly on technology, partly on physiology and partly on the attitudes of diving companies, operators and regulatory bodies.

Improved equipment and gas mixes are enabling step-by-step increases in diving depths, but as they do so the physiological difficulties become increasingly severe, and a limit upon working depths must inevitably be reached. The influence of attitude is demonstrated by the differing diving limits which apply in different parts of the world. Typical limits for different classes of diving are shown in Fig. 1.

By the above definition, very few deepwater flowline tie-ins have been performed. A larger number of tie-ins have been performed using equipment suitable for deep water, but usually with divers present or at least with the knowledge that they could be called in if necessary. This is partly due to understandable caution about using new technology without any back-up and partly because, by the time diverless techniques are developed and proven, advances in diving have caught up with the previously unreachable water depths. However, the situation is changing as confidence grows in diverless techniques and as the inevitable physiological limits are approached.

The definition of deep water can be further refined to that of very deep water: that is water beyond the design depths of existing equipment. Broadly speaking, this starts at twice the depth of deep water (currently 500–600 m).

At such depths the limitations are again physical, but in this instance they relate to the equipment, in particular umbilicals and control systems.

EXISTING PRACTICE

The majority of existing flowline tie-ins employ spoolpieces. The two principal advantages of spoolpieces are that:

- they can absorb some misalignment of the flowline
- they absorb longitudinal expansion of the flowline

Spoolpieces may be either flexible pipe of composite construction or rigid steel. In the case of rigid steel spoolpieces, careful measure-

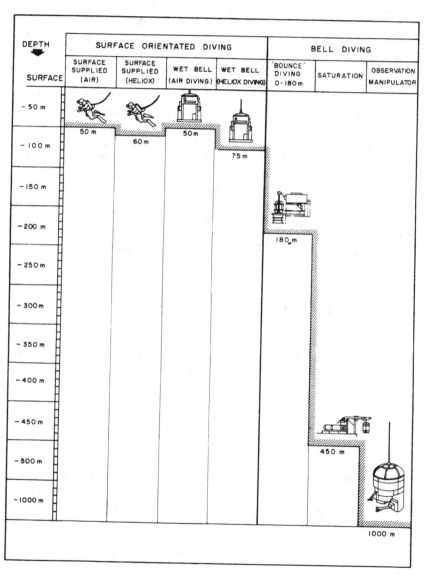

Fig. 1 Typical diving depth limitations

ment is necessary on the sea bed to ensure that the spoolpiece is made to the required dimensions.

Connection (i.e. final mating of the flowline to the spoolpiece or the spoolpiece to the facility) is commonly by bolted flanges or by hyperbaric welding. Both of these methods are currently dependent on divers. While some proprietary mechanical connectors are, in principle, less dependent on divers, they have mostly been installed with diver assistance.

Where the flowline is to be tied in to a fixed structure, the need for a subsea tie-in can be eliminated by pulling the flowline through a J-tube to the surface (Walker and Davies, 1983). The method is, of course, only available when a suitable J-tube can be made available on the structure. It is further restricted in that it can, in general, only be used at the first end of the flowline to be connected, a limitation that is shared by a number of the diverless tie-in methods which are discussed in more detail later.

Where flowlines have been installed in bundled form, the lines have generally been tied in individually. J-tube pulls of strapped bundles have been performed, and tests at various scales have proved the feasibility of the method for cased bundles (Maten, 1985).

FLOWLINE INSTALLATION

Before considering connections, it is useful to look at the different forms of flowlines and installation methods, as both have an important bearing on the selection of tie-in method.

Flowlines

Flowlines may be categorized by construction, as either flexible or rigid. Flexible flowlines are typically of a complex construction with several layers of elastomers and metal designed to retain pressure, prevent collapse, minimize gas permeation, prevent corrosion, resist axial and bending load, and permit the required amount of movement. Rigid flowlines are similar to other small-diameter steel pipelines. On occasion, specialist materials are used to combat the corrosive effects of the wellstream fluids.

The second way in which flowlines may be classified is as single lines, or as lines installed as part of a bundle. It is often more economic to install flowlines and umbilicals bundled together rather than as single lines. Bundled lines also lead to less congestion on the sea bed. Bundles can range from a control umbilical strapped to a larger pipeline in piggy-back fashion to complex assemblies where a carrier pipe contains multiple flowlines, injection lines and umbilicals.

Installation Methods

Four main installation methods can be identified:

- S-lay
- J-lay
- reel
- tow

The applicability of these methods to flowline types is illustrated in Table 1 and their depth ranges are shown on Fig. 2.

TABLE 1
Applications of Installation Methods

FLOWLINE TYPE	CONFIGURATION	INSTALLATION METHOD			
		S-lay	J-lay	Reel	Tow
Flexible	Single			Size limited by Manu-facture	Unproven
	Bundle	As piggy-back line	As piggy-back line	Under Develop-ment	Unproven
Rigid	Single	Size limited by tensioners and mooring	Requires Develop-ment	Size limited by reel	1st deep water project in engineering
	Bundle	Limited Applicability		Under Develop-ment	Approx. 10 Installations by Mid-depth Tow

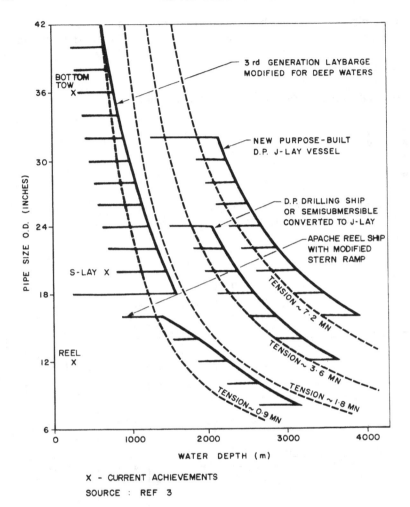

Fig. 2   Estimated maximum depth capabilities of various pipelay systems

The methods are discussed in more detail below.

S-lay

This is the classic method of installing subsea pipelines. Individual pipe sections are welded together on a special-purpose vessel and lowered, under tension, to the sea bed in a continuous process.

In order to lay large diameter pipe in deep water, extremely high tensions would be required. Conventional anchor systems also impose depth limitations. Laybarges are highly optimized vessels; it would be difficult to incorporate the additional facilities necessary to fabricate complex bundles and, even if it were possible, the production rate would be greatly reduced. Small control or injection lines have, however, been strapped to large diameter pipelines and laid in this way.

J-lay

This method is currently being considered to overcome the depth limitations of S-lay. In this method, the pipe leaves the barge vertically rather than horizontally, thus reducing the tension requirements and eliminating the "stinger" tail ramp (Langer and Ayers, 1985). The practical use of the method depends on the implementation of a rapid single-station welding process (de Sivery *et al.* 1980; Turner, 1985).

Reel

In this method a continuous length of single or bundled pipe is unwound from a reel mounted on a vessel, and lowered to the sea bed. Where the pipe is rigid, it must be deformed plastically to wind it onto the reel and again to straighten it for laying. As no welding is required, quite high lay rates are achievable, even including the fitting of anodes. While umbilicals can be piggy-backed to reeled pipelines, the use of reeling has been largely, although not entirely (Noroil, 1985), confined to single pipelines.

Development work is being carried out to extend the method to rigid flowline bundles. One major difficulty with this in the past has been to devise a way of simultaneously straightening the pipes making up a rigid bundle. A current proposal overcomes this by twisting the bundle into a spiral form (*Offshore Engineer,* 1986a).

Tow

In this family of methods, a pipe string is fabricated onshore and towed to the required location by one or more tugs. Unlike the other methods discussed, tow is frequently employed for bundled pipelines.

Tow methods are usually subdivided according to the towing depth of the bundle. The categories are as follows:

- bottom tow
- off-bottom tow
- mid-depth tow
- near-surface tow
- surface tow

Of these, the last is only suitable where environmental loads during installation are very low.

Apart from the bottom tow of short sections, all the methods normally depend on the provision of extra buoyancy to offset the weight of the pipe. This may take the form of buoys, buoyancy tanks, foam or a carrier pipe enclosing the bundle. As the water depth increases, the provision of buoyancy becomes more difficult because the water pressure increases its cost and complexity.

To date, tow methods have been limited to bundles less than 4.5 km long (*Offshore*, 1986).

FACTORS AFFECTING TIE-IN

Site Connection Conditions

Typically, the flowlines will be tied in to different types of structure at either end. For example, they may run between a satellite wellhead and a subsea production system or between a subsea production system and a fixed platform. The nature of the structure can significantly affect the tie-in method selected.

For a fixed structure, there is likely to be considerable congestion in the platform approaches, and therefore solutions with minimum space requirements are likely to be favoured (e.g. bundled flowlines rather than single ones). Single wellheads typically have two flowlines plus umbilicals; therefore the sea bed is relatively uncluttered and tie-in methods can be selected without concern over space constraints.

Subsea production systems frequently suffer from similar congestion problems to those of platforms, problems that can be aggravated by the sweep space required for lateral deflection based tie-in methods.

Effect of Water Depth

Water depth has major implications for equipment as well as for divers. A lot of current tie-in equipment is not suitable for use in deep water despite being nominally diverless, depth limitations arising from a number of causes.

Where control and/or power is delivered by an umbilical, depth is obviously limited by the length of the umbilical. Lengthening the umbilical may not be a practical solution since increased losses can mean that control signals are not reliably received and power losses can lower the effectiveness of machinery. A radical redesign of the umbilical may be required, and possibly also of the control and power systems.

Mechanical control from the surface also becomes more problematic at greater depths. The load imposed by the self weight of cables increases and they become "spongier" due to increased length. Environmental loads cause larger movements, increasing the risk of tangled cables. At depths beyond 600 m it is likely that major redesign will be necessary for some mechanical control systems.

A further limiting factor can be pressure resistance of housings and the efficiency of buoyancy tanks.

While control and power problems can be minimized by locating as much equipment as possible on the sea bed, this increases the exposure to high pressures, increasing the need for complex housings seals, pressure compensators, etc.

TIE-IN METHODS

It might be natural to assume that each pipeline has two tie-ins, one at each end, but this is not always the case. Where the installation method is only suitable for limited lengths of flowline, midline tie-ins may be required.

Similarly, it is convenient to divide tie-ins according to the sequence in which they occur, i.e. first end and second end. However, there is one installation and tie-in method where both ends are tied-in at the same time. Methods suitable for the various tie-in positions are shown in Table 2, and discussed further below.

First End Tie-ins

The factor classifying a first end tie-in is that the pipeline is free to move axially over a relatively large distance, making a wider range of methods available.

Where the pipeline is being laid (S-lay, J-lay or reeling), only a short length need be placed on the sea bed, with the installation vessel standing by until the tie-in is complete: this means that the

TABLE 2
Tie-in Methods

Tie-in position	Method	Comments	Field experience	Confidence limits
First end	J-tube	Requires fixed platform; not proven for cased bundles	14-inch in 360 m	600 m
	Connect and lay away	Requires swivel and strong tie-in structure	flexible in 300 m	600 m
	Direct pull (single wire)	Simple; has been widely used	Wellhead connections only	450–1000 m
First or second end	Spoolpiece	Flexible, articulated or sprung spoolpiece assists alignment tolerances	hyperbaric welds to 300 m (welding proven to 360 m)	600 m with guidelines
	Lateral pull	Requires a number of control wires and sheave blocks	diverless in 80 m	600 m
Mid-line	Lift and connect	Not proven in deep water; could be used for buoyant bundles	90 m	–
Simultaneous	Simultaneous draw down	Has been installed in scaled-down form only	100 m	300 m +

force needed to manoeuvre the flowline can be comparatively small. Where a tow method is used the forces may be limited by the provision of buoyancy (Fig. 3).

The other main advantage of first end tie-ins lies in the fact that the flowline or flowline bundle can be placed directly in front of the tie-in structure at a convenient distance away and pulled-in in a straight line.

A variant on this technique is to make the connection before any pipe has been laid on the sea bed, and then to lay away, effectively using the tie-in structure as a lay initiation anchor.

Second End Tie-ins

For second end tie-ins, a direct pull is not generally possible; first because the entire pipeline is in contact with, and restrained by, the sea bed (except for buoyant-tow installation methods), and secondly because the length of the flowline has to be compatible with the distance to the tie-in structure.

Two basic methods are available for dealing with the length movement allowance and tolerances:

Spoolpiece Tie-in

The spoolpiece method has to be adapted from the techniques employed in shallower water. Hyperbaric welding is not currently available without diver intervention, so some form of mechanical connector is required. Axial movement of both ends of the spool has to be available if an in-line connector is used; this also permits misalignment to be taken up.

Sufficient flexibility can be obtained in the spoolpiece by:

- using flexible pipe
- using a sprung spool (Fig. 4)
- incorporating swivel joints (Fig. 5)

Lateral Deflection

Lateral deflection involves positioning the flowline end to one side of the target structure and then pulling it laterally into position (Fig. 6). This has two disadvantages compared with a direct pull-in.

- alignment is more difficult to achieve
- a clear (swept) area is required to one side of the tie-in site

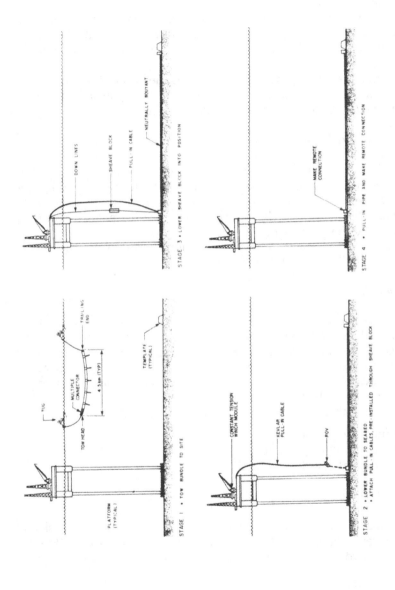

Fig. 3   Direct pull-in for towed pipe (first end connection)

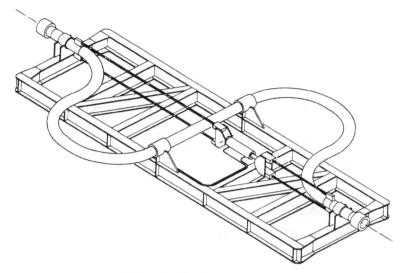

Fig. 4 Flexible spoolpiece

The flowline may be pulled towards the target by a single wire, or a series of wires may be deployed through dead-man anchors to give greater control of alignment. Final alignment is usually assisted by a bell mouth or stab-in guides.

For large diameters such as export lines or bundles, it is necessary to make a length of the pipeline neutrally buoyant. This gives greater flexibility and reduces the pull forces, but can expose the pipe to large current forces.

One development of this technique is the use of vertical deflection rather than lateral deflection. The required initial shape could be attained by local adjustments to buoyancy, pull-in being again by a system of wires. The principal advantage of this method is that it does not require the same amount of seabed space.

In addition, it should be possible to devise initial configurations which it would be difficult to create laterally by laying or towed installation. Direct pull for second end tie-ins may then become available by creating a vertical slack loop behind the pullhead.

A second possible development area lies in the use of subsea vehicles to perform the tie-in. Controlling operations by wires from the surface becomes increasingly difficult as water depth increases; there is a significant risk of lines becoming tangled. A tracked vehicle with power and control signals supplied by a suitable umbilical would be less depth dependent and could carry equipment for making the connection as well as for the pull in (Fig. 7).

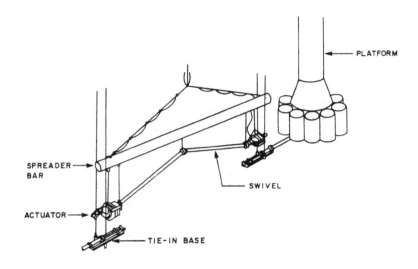

- INSTALL ARTICULATED SPOOL INTO POSITION ON TIE-IN BASES

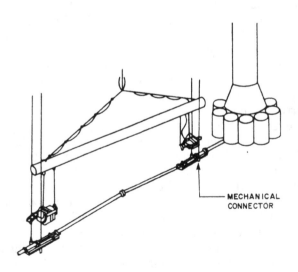

- MAKE CONNECTION
- REMOVE SPREADER BAR AND ACTUATORS

Fig. 5   Articulated spool (first and second end connections)

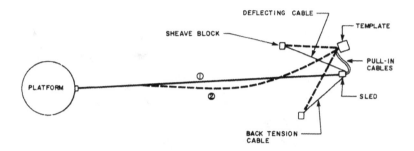

- DEFLECT BUNDLE
- PULL-IN DEFLECTING CABLE AND LET-OUT BACK TENSION CABLE TO CONTROL DEFLECTED SHAPE
- BUNDLE MOVES FROM ① TO ②

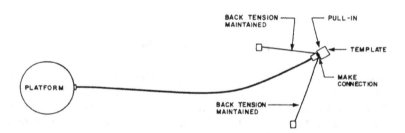

- PULL-IN SLED FROM TEMPLATE TO MAKE CONNECTION
- RELEASE BUOYANCY TANKS AND FLOOD BUNDLE

Fig. 6 Horizontal deflection (second end connection)

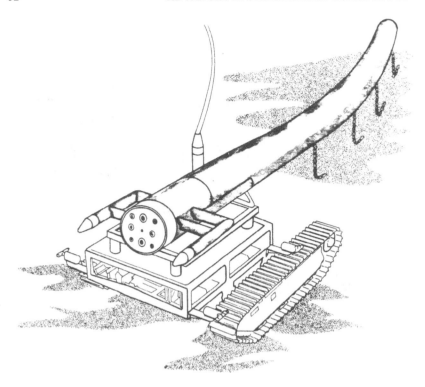

Fig. 7   Seabed tie-in vehicle

Another way of eliminating pull wire connections to the surface is to employ subsea winches. The seabed configuration remains the same as for a conventional lateral tie in, but only a control and power umbilical connects the system to the surface. A typical subsea winch, developed by ACB, is shown in Fig. 8.

Mid-line Tie-ins

Mid-line tie-ins can be performed in the same way as first or second end tie-ins, but because they are not located adjacent to a fixed seabed installation, they can also be performed in theory by lifting the flowlines to the surface and connecting them there; this obviously allows a wide range of connection methods to be used, including welding.

The method has been limited to comparatively shallow water (not more than 90 m) in the past. This is because the davit lifting

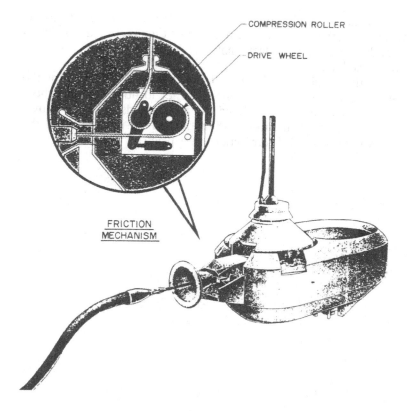

COMPRESSION ROLLER

DRIVE WHEEL

FRICTION
MECHANISM

Fig. 8 Subsea winch

techniques employed to lift the two pipe ends in a S-curve rapidly
become impractical at greater depths. However, it is much easier to
lift the flowline ends in a catenary. A curved spoolpiece may then be
welded into place and the completed flowline laid laterally on the sea
bed. The cost implications of the extra length of line would have to be
offset against the cost of a subsea tie-in.

Where a mid-line tie-in is present, both ends of the flowline may be
tied-in using "first end" methods.

Simultaneous Connection

There is a proprietary system of flowline installation which employs connection at both ends simultaneously (Coleman *et al.*, 1982); this is known as simultaneous drawdown. Perhaps the greatest attraction of the method lies in the fact that the connection operation also positions the bundle.

The method is illustrated in Fig. 9. The bundle is towed out by the near-surface tow method. When the target locations are reached, drawdown cables are attached and the bundle ends are pulled down towards the tie-in structures. When the bundle is in position the buoyancy is released and connection takes place.

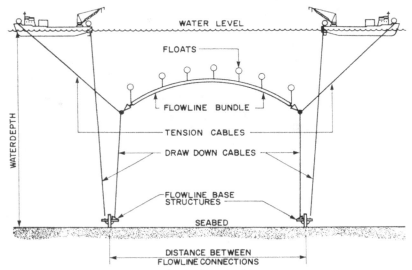

Fig. 9   Simultaneous drawdown method

ALIGNMENT

Once the flowline has been pulled into position it is necessary to ensure that the connector faces are in alignment. In some cases the connector systems have a facility for accepting or correcting some misalignment; in other cases the pull-in frame has to establish precise orientation by means of stab-in guides or similar features. It is considerably easier to establish the required alignment for a flexible flowline than for a rigid one.

Bundles provide an extra level of difficulty as twisting of the bundle will prevent correct alignment of multi-bore connectors.

If buoyancy is provided, a greater length of pipeline may be moved, so reducing alignment forces.

FLOWLINE CONNECTORS

The use of mechanical connectors for pipelines and flowlines is fairly well established, with the total number of installations running into hundreds. Almost all installations have involved the use of divers, most have been of a single pipeline, and many have involved flexible lines.

The technology has borrowed heavily from wellhead design which does not necessarily have the same requirements as flowline design.

The main design goals for deepwater flowline connectors may be stated as follows:

- The connectors must achieve a high reliability metal-to-metal seal, with the seal protected against damage during hub mating
- the system should have a strength and integrity comparable with that of the flowline
- the integrity of seals should be capable of verification before system commissioning
- defective seals should be retrievable
- hydraulic or other equipment necessary to "unmake" the connection should not be left subsea
- the system should be capable of absorbing some misalignment of the flowline
- a multibore connector should be available for bundles

Not all these requirements are appropriate for all applications and some of them work against each other. For example, maximum integrity requires the simplest system with the fewest components, but misalignment devices inevitably add complexity and an additional potential leak path.

Many systems feature external pressure-testing ports. While an external pressure test does not give complete assurance of sealing against internal pressure, it is nonetheless useful.

Some of the systems offering multibore connectors (McEvoy, Vetco) feature removeable seal plates. Where this is not the case (e.g. the Cameron multibore connector) the connector can be incorporated in a spoolpiece; the use of a spoolpiece naturally requires two connectors per tie-in.

Once a connection has been successfully made, tested and commissioned, experience suggests that there is little risk of subsequent failure. Most experience is, however, fairly recently acquired; it is not prudent therefore to assume that trouble-free operation is guaranteed for long design lives. Where a long design life is required, it is desirable that all actuating equipment should also be retrievable to the surface, in addition to siting the connector away from areas of high bending moment or fatigue loading.

A connection system which is capable of accommodating misalignment allows a less sophisticated pull-in system to be employed. Two types of misalignment are generally considered; axial and angular.

Axial misalignment may be taken up by a slip joint (BIMS) or by a series of ball joints. In either case, the use of a spoolpiece is implied. However, this spoolpiece may be mounted within the template piping arrangement and used to accommodate expansions. Angular misalignment may be taken up by a single balljoint.

Cameron have recently developed a version of their collet connector which is capable of accepting a misalignment of up to $\pm 3°$. There are no misalignment systems for bundled flowlines. The need to absorb misalignment is one reason why flexible pipe has been widely used, either as a spoolpiece or a complete flowline.

While bundled flowline connectors exist (e.g. McEvoy, Cameron), experience in the field is limited. A number of systems are currently under development (e.g. for Elf's Skuld, Shell's DIMOS, and a diverless connector being developed for Statoil by Liaaen Engineering). Confidence limits of up to 900 m may be placed on some existing connector systems, but field experience is limited to 300 m.

AN ASSESSMENT OF TECHNIQUES AND REQUIREMENTS

Table 3 presents a checklist of items to be considered in selecting a tie-in method and connection system.

Despite the uncertainty provoked by recent price upheavals, there is little doubt that there will be a continuing trend towards subsea completions, and that many of these will continue to take place in deeper and deeper water (e.g. *Offshore Engineer*, 1986b). The first consequence of this is likely to be a move towards more diverless systems. Many of these now exist although few have been used in true diverless conditions.

The cautious incremental design, implementation, and evaluation of diverless components which has gone on over the past decade has brought with it an increasing confidence. A number of schemes have been sponsored by oil companies (Coleman *et al.*, 1982; Pras *et al.*, 1978; Renard *et al.*, 1983; Sinclair *et al.*, 1976; King *et al.*, 1983;

TABLE 3
Tie-in and Connector Checklist

	External factors	Features
Tie-in methods	No. of flowlines Size, service and length of flowlines Bundled or single configuration Rigid or flexible Tie-in sequence (1st end, 2nd end etc.) Available seabed space Water depth Flowline expansion forces movement Alignment requires of connector	Diver dependence Subsea equipment requirements Vessel requirements Space requirements Track record Ability to absorb expansion Cost
Connection Systems	Single or bundled flowlines Presence of spoolpiece Movement available to make connection Water depth Loads applied by flowlines	Multibore capability Seal replacement Retrievable hydraulics Reversibility Track record Cost

Thiberge, 1986) partly as development exercises. It is likely that some of these will be deployed on a more commercial basis within the next few years.

For very deep water, further technological developments are likely to be required. Wire-based control from the surface becomes increasingly impractical at depths beyond 600 m. Either the technology must be modified (e.g. different types of wire or increased spacing) or more control has to be moved to the sea floor. This may take the form of free-swimming (Fig. 10) or seabed vehicles or the placing of winches, etc. on the sea bed.

Even where existing equipment is basically suitable for very deep work, it will normally require adaptation in terms of longer umbilicals, increased pressure resistance, etc. There is also scope for more automation in tie-ins with surface computing systems monitoring the progress of operations, suggesting means of overcoming problems and warning of imminent dangers.

The loads generated by different tie-in methods vary considerably. This makes some methods unsuitable where the tie-in is to a satellite well with limited load resistance. While a template or manifold centre may have greater strength, it is important that tie-in methods and

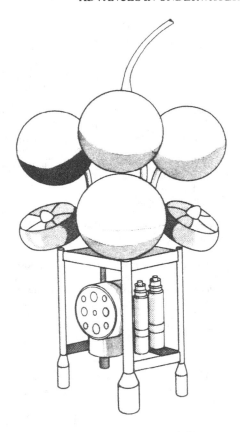

Fig. 10   Free-swimming vehicle carrying flowline connector tool

loads are considered at an early stage in design. Operational expansion of the flowlines can also impose severe loads on adjacent structures. The conventional means of dealing with these loads is by installation of an expansion offset, though this limits the choice of tie-in methods.

The use of bundled installation is likely to increase. In order to take full advantage of this, increasing use is likely to be made of multibore connectors. A number of these exist or are under development, but field experience is fairly limited to date.

The concept of a flexible spoolpiece is attractive (not least because it could absorb expansion movements of the bundle), but accurate orientation of the bundle would still be required.

THE NEED FOR FURTHER DEVELOPMENTS

At first sight, the designer seems to have an enviable choice of flowline types and methods available for installation, tie-in and connection. Closer examination shows this is not the case. Only certain combinations of installation and tie-in methods are practical; similar restrictions exist for combinations of tie-in and connection methods, while the choice of a flowline type can impose further restrictions.

Of the technically feasible options, only a limited number have been developed to the stage of manufacture of hardware and not all of these have been installed. Of those which have been installed, few have been put in without divers. Further restrictions may arise out of proprietary rights to designs or items of equipment. The choices for designers wishing to use field-proven diverless equipment are thus quite meagre.

The need for further development work has been recognized by more than one oil company, and several development programmes are underway. The development cycles for subsea systems have tended to be quite long; while this is partly due to caution and the availability of proven alternatives, installation methods should be considered at the earliest possible stage of field development. The following design stages can be identified:

- concept evaluation
- concept development
- prototype development
- field testing
- final development and design
- installation

Concept Evaluation

The different methods and components are compared in terms of suitability for the proposed field development, availability and development effort.

Concept Development

The best method(s) identified at the evaluation stage are developed to the level of preliminary design. Mathematical modelling and analysis of the system behaviour is performed, possibly linked to a programme of model testing. Installation procedures are written.

Detailed cost and schedule estimates are developed.

The first two stages are relatively inexpensive and can reap large rewards by ensuring that the detailed design proceeds on a rational basis, that simple economic options are not overlooked, and objectives for further work can be met. Subsequent stages are more expensive and require careful justification.

Prototype Development

A full-size or reduced-scale prototype of the proposed system is built. Workshop tests establish the functioning, strength and reliability of the components.

Field Testing

The complete system is installed in realistic conditions. This is often done in a production system, but within the reach of divers. The performance of the system is carefully monitored.

Final Development and Design

Lessons learned from the prototype are incorporated into a revised system.

Installation

The system is installed in a deepwater location.

In order to obtain the maximum benefit from a programme of this type, the design should contain as many standard components as possible.

AREAS TO WATCH

As has already been pointed out, the number of flowline tie-in methods which could be developed is very much greater than the number which have been developed; it is likely that this will continue to be the case. However, there are some areas where developments have sufficient potential to merit the expenditure of the sums required. While prediction is always a very risky business, we suggest that the following will be areas worthy of consideration:

- bundle tie-ins
- subsea vehicles
- the application of expert systems to real-time control of mating operations

REFERENCES

Coleman, J. T., Saliger, K. C. and Yancey, I. R. 1982. Method of laying and connecting intrafield flowline bundles in deep water. *Offshore Technology Conference*. Paper OTC 4269.

Langer, C. G. and Ayers, R. R. 1985. The feasibility of laying pipelines in deep waters. *ASME Offshore Technology Conference*. Paper OTC 4870.

King, A. R., Ostocke, H. D. and Best, M. J. A. 1983. Development of a diverless subsea flowline and control-line connection system. *Offshore Europe 83*, September, pp. 501-511.

Maten, G. J. 1985. Troll field flowline bundle J-tube pull-in concept studies. *4th Int. Offshore Mechanics & Arctic Engineering Symposium*, February, vol. I, pp. 671-675.

Noroil 1985. *Subsea Technology* 13 (5, May), 29-62.

Offshore 1986. Major subsea project starts at Scapa. August, p. 82.

Offshore Engineer 1986a. Bundles offer real advantages. March, pp. 39-42.

Offshore Engineer 1986b. TOGI pushes subsea gas frontiers for extra oil. October, pp. 26-27.

Pras, S., Levallois, E. and Gouraud, O. 1978. Second end flowline connection without length adjustment. *Offshore Technology Conference*. Paper OTC 3074.

Renard, B., Lerique, M., Tinchon, J., Muller, D. and Gosselin, J. 1983. Specific features of a new concept for deepwater flow-line laying and connecting and its sea trial in 250 meters. *Offshore Technology Conference*. Paper OTC 4577.

Sinclair, A. R., Burkhardt, J. A. and Daughtry, A. C. 1976. Deepwater pipeline connections – a subsystem of the submerged production system. *Offshore Technology Conference*. Paper OTC 2526.

de Sivry, B., Sudreau, B., Carsac, C. and Jeqousse, M. 1980. Electron beam welding of J-curve pipelines. *Offshore Technology Conference*. Paper OTC 3746.

Thiberge, P. 1986. Workshop and shallow water tests of a new sealine mechanical connector. PI no. 1619, January.

Turner, D. L. 1985. Flash butt welding of marine pipelines today and tomorrow. *Offshore Technology Conference*. Paper OTC 4870.

Walker, A. C. and Davies, P. 1983. A design basis for the J-tube method of riser installation. *ASME 2nd Offshore Mechanics & Arctic Engineering Symposium*.

7

Intervention for Maintenance

W. E. Carter, P. E. Hadfield and P. Metcalf,
Furness Underwater Engineering Ltd

INTRODUCTION

The ability to maintain any subsea production system is the cornerstone to its success.

Careful design of the system and its components may reduce the number of areas requiring maintenance; better component reliability may make maintenance less frequent; built-in redundancy may make it possible to postpone maintenance once a failure occurs and continue production, but at some stage during the life of a subsea production system, maintenance *will* be required.

Maintenance of equipment in shallow water is by now well established; it uses techniques and equipment in every-day use on land, with only minor modifications for subsea use.

Lifting and positioning of equipment by the use of cranes and winches, cleaning and inspection, cutting and welding, and the use of hand and hydraulic tools by divers are all routine and are carried out within environmental limits much as they would be on land. Production equipment itself has not changed significantly from the surface equipment from which it was derived.

Advances in Underwater Technology, Ocean Science and Offshore Engineering, Volume 10: Modular Subsea Production Systems
© Society for Underwater Technology (Graham & Trotman, 1987)

Even the introduction of Remotely Operated Vehicles (ROVs) has had a very limited effect on the design and maintenance philosophy of shallow-water production equipment, ROVs being used principally for observation and inspection, diver back-up and simple manipulative tasks.

The advent of the need to conduct maintenance in deep water does, however, require a radical new approach: land-based techniques and equipment cannot be adapted successfully for use beyond diver depths, so the new techniques must be remotely operated from the surface. These new techniques will bring benefits to shallow-water production as they are potentially more cost-effective maintenance techniques than current diver intensive methods.

It is important that sufficient time and attention be given to these new techniques at the early stages of any project as they have a major effect on production system layout, equipment design and operating philosophy. Better identification and planning for both routine and potential maintenance requirements is needed to compensate for the dexterity of divers and their ability to improvize.

Alternative means of carrying out specific tasks must be designed in, since modification to a production system installed in deep water will be difficult.

Since these remotely operated intervention techniques are largely new, they must be thoroughly tried and tested, preferably early in the project, and certainly before installation.

Finally, the overall cost of maintenance and intervention must be considered, and the most cost-effective combination of capital, development and continuing expenditure planned for.

MAINTENANCE/INTERVENTION PHILOSOPHIES

It is becoming apparent that some common philosophies for the maintenance/intervention of deepwater subsea production systems are being adopted throughout the industry.

However, the technical solutions to these common problems that have been seen to date vary enormously in concept. This is probably due to the fact that each development is unique: in terms of production size (number of wells, satellite completions, water injection, etc.), in location (water depth, proximity to existing developments and supply bases) and in environment (wind and wave spectra and current profiles). Furthermore, each operating company has preferred operating techniques and will develop along the lines of current in-house research and development programmes.

It is only when deepwater subsea developments have been thoroughly exercised under real operating conditions that any clearly winning solutions will emerge. These are inlikely to be on a "system" level, but in local areas such as valve insert or control pod design and replacement tooling. Should such solutions become industry standards, costs should reduce and maintenance equipment become available "off the shelf".

Philosophies being commonly adopted are as follows:

- *Elimination of manned intervention.* Below diver depths this includes a reluctance to use atmospheric diving systems (ADS), manned submersibles, dry transfer systems and possibly submarine systems.
- *Limited in-situ repairs.* In-situ repairs are limited to equipment that cannot sensibly be retrieved, e.g. pipelines and structures.
- *Widespread repairs by replacement or retrieval for surface repairs.* Due to the nature of subsea production equipment, comprising electronics, hydraulics and close tolerance mechanical components, most equipment will be replaced or retrieved for surface repair.
- *Modularization.* Some form of modularization is required to enable retrieval of failed components by the intervention equipment chosen. Larger modules imply less connections and seal faces; a traditional source of problems, but increase handling difficulties.
- *Back-up methods of intervention.* The intervention techniques adopted must not be allowed to make a case of failure more costly if they themselves fail. Obviously reliability of tooling and technique will have been tested and proven, but alternative methods must be available as a precaution. An example is the mechanical override of connectors by ROV as a back-up to hydraulic operation by remote tooling.
- *Simplicity and ruggedness.* Experience with the Shell UMC Remote Maintenance Vehicle has shown that sophisticated intervention equipment can work successfully, but that simpler solutions may be more appropriate and easier to operate and maintain. The necessity for equipment to be rugged should not need to be pointed out, but it is sometimes overlooked.

INTERVENTION TASKS

Intervention tasks will differ for each subsea production system, but they generally fall into one of the following groups based on relative complexity, size/weight of equipment or modules, and associated surface support requirements.

Group 1 – Inspection and Surveillance

This first group consists of purely observational work such as regular surveys or support of other activities that can be carried out by an ROV from a DP monohull vessel. Typical activities include:

- inspection for overall damage, leakage, marine growth, seabed scour and build-up
- confirmation of valve operation by observation of status indicator
- observation of module landing/retrieval

Group 2 – Manipulative Tasks and Lightweight Module Replacement

This group consists of all light work tasks and back-up requirements that can be carried out by an ROV with existing manipulators and tooling or specially designed tooling packages. An approximate weight limit for modules is 2 tons. Tasks can be carried out by an ROV from a DP monohull vessel. Typical activities include:

- cleaning, jetting, NDT activities
- minor debris clearance
- valve actuator or connector override
- seal plate replacement
- valve insert replacement
- control pod replacement

Group 3 – Heavy Module Retrieval/replacement

This group consists of the retrieval/replacement of modules to the limits of lifting capability of a monohull or small semi-submersible support vessel, i.e. module weights in the range 2-40 tons: the means of lifting is either by liftwire or by drillstring. Typical tasks depend entirely on the modularization philosophy adopted, but exclude modules requiring riser connections and controls such as Xmas trees or BOP stacks.

Group 4 – Well Servicing

This group includes any activity that requires the services of a drilling rig or ship and riser connections and controls. Typical activities include:

- well workovers using wireline
- Xmas tree retrieval/replacement
- BOP stack running/retrieval

The requirement for this form of intervention may be reduced by through flowline (TFL) techniques.

Group 5 – Major Damage Repair

This final group consists of the retrieval/replacement of major structural modules. This type of intervention is equivalent in scope to the installation process, and would require the services of a crane barge and drilling vessel, etc. Typical module weights are in the range 40–800 tons. The aim of the design team must be to identify all maintenance requirements and possible failure modes, predict their frequency and their consequences (lost production, pollution, cost to repair) and plan maintenance activities accordingly.

Where the effect of failure is serious, redundancy must be built into the system to allow continued production or prevent pollution. Where failures can be predicted with some certainty, preventative maintenance can be planned.

INTERVENTION SYSTEMS

Intervention systems consist of a number of elements which must perform well individually but, more importantly, together as a system. This can only be ensured by good systems engineering followed by adequate testing. The following are the major components of an intervention system:

- surface support vessel
- deployment/handling system
- in-water transportation
- guidance system
- tooling
- back-up system

Some of the options for each of these are discussed below.

Surface Support Vessel

Options include monohull DSV; monohull drillship; semi-submersible vessel; semi-submersible drillship; floating production unit; and crane barge.

Technical requirements include sufficient deck space/handling capacity/adaptability for different tasks; station keeping; and stability. Significant advances have recently been made in DSV design to make them more suitable for subsea maintenance.

Commercial requirements include availability when needed; fast transit time from the suply base; but, above all, cost-effectiveness for the task in hand.

Deployment/handling Systems

Options are limited by the choice of surface support vessel and the level of its on-board equipment.

Deployment may be overstern, overside or through a moonpool. Overstern deployment accommodates heavy loads without loss of stability, but is vulnerably close to propellers and thrusters and may suffer excessive vertical movement from pitch. Overside deployment may be limited by vessel stability and suffer from vertical movement from roll. Moonpool deployment gives the greatest protection to equipment and least vertical movement as it is close to the metacentre, but may limit equipment size.

Handling may be by crane, "A" frame or special gantry. The gantry gives the greatest control, but is possibly the most expensive, heaviest and least flexible option. A cursor or guideframe system may be used to assist transfer through the air/water interface if necessary.

Whatever the choice of deployment/handling system, they will generally be simpler and more effective if permanently installed than if transferred as needed between vessels of opportunity.

In-water Transportation

Options are to suspend the equipment/tooling on a liftwire or drillstring, or attach it to a free-swimming or tethered vehicle with appropriate buoyancy. The choice is usually determined by the mode of access available at the seabed installation; either vertical or horizontal.

With the liftwire or drillstring access must be vertical and unimpeded, and some form of heave compensation system is needed to decouple the equipment from the rise and fall of the surface vessel as shown in Fig. 1.

With the free-swimming or tethered vehicles, which may be manned submersibles or submarines, atmospheric diving systems such as bells or suits, ROVs or special purpose vehicles, access may

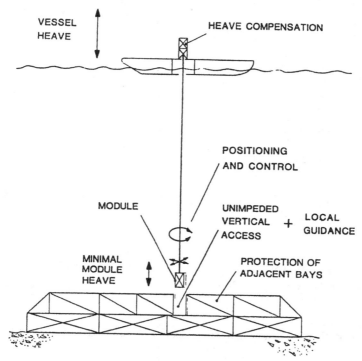

Fig. 1 Requirements for liftwire or drillstring in-water transportation

be vertical *or* horizontal. Furthermore decoupling is inherently achieved, a small surface vessel may be suitable, and the vehicle may already be on hand for other tasks. However, their requirement for neutral buoyancy does impose limits on the size of equipment/tooling carried, and they do not cope easily with variable loads at different stages of the intervention activity, e.g. a vehicle trimmed to install a heavy module will be strongly positively buoyant on its return trip to the surface. Solutions to this problem are either to introduce variable buoyancy by blown chambers or pumped ballast, to install weight compensators that are picked up by the vehicle in place of the module as shown in Fig. 2, or to use a single trip tool to carry down a replacement and return with the faulty equipment as shown in Fig. 3.

Guidance Systems

The guidance system must be able to position and orientate the

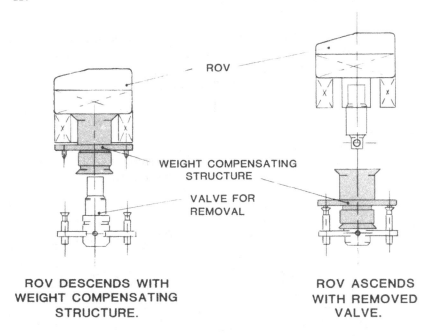

ROV

WEIGHT COMPENSATING
STRUCTURE

VALVE FOR
REMOVAL

**ROV DESCENDS WITH
WEIGHT COMPENSATING
STRUCTURE.**

**ROV ASCENDS
WITH REMOVED
VALVE.**

Fig. 2   Use of a weight compensating structure to trim a neutrally
buoyant vehicle

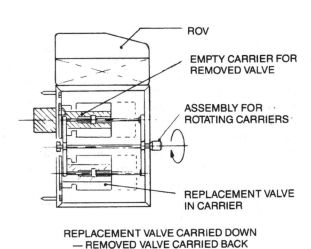

ROV

EMPTY CARRIER FOR
REMOVED VALVE

ASSEMBLY FOR
ROTATING CARRIERS

REPLACEMENT VALVE
IN CARRIER

REPLACEMENT VALVE CARRIED DOWN
— REMOVED VALVE CARRIED BACK

Fig. 3   Use of a single trip tool to trim a neutrally buoyant vehicle

equipment with ever-increasing accuracy the closer it approaches the subsea installation.

The surface vessel must be kept in a constant position relative to the template using surface, satellite or subsea navigation techniques coupled to DP, or alternatively by anchoring if the vessel is to be on station long term. The mean position may be well away from the template due to currents.

On arrival at the template the equipment must be captured by some means to enable it to be decoupled from surface movement prior to docking within the final capture zone of the connectors or tools.

When running equipment vertically on liftwires, the use of guidelines/guideposts for guidance is well established, as shown in Fig. 4.

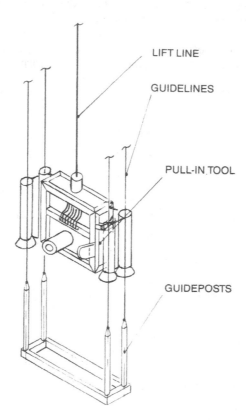

Fig. 4   Use of guidelines/guideposts to run a typical pull-in tool

Guidelineless techniques using the surface vessel to position the equipment are well proven for running drilling equipment into isolated wells, but not into crowded templates where adequate protection for surrounding buoys must be provided.

A simple thruster pack at the end of the liftwire may be used to decouple the equipment from surface-vessel drift and improve positioning ability. The equipment may be captured by guidance structures based on interlocking forms that correct any initial position error as the modules are lowered to their final position. These may also be used with equipment being run vertically beneath free-swimming or tethered vehicles. These principles are illustrated in Fig. 5.

A further alternative is a hauldown system as used for the Remote Maintenance Vehicle (RMV) on the Central Cormorant Underwater Manifold Centre, shown in Fig. 6. The RMV is attached to a line released from the UMC and winches itself down towards the template.

A number of methods can be used to position equipment horizontally. Docking and alignment can be performed by extended

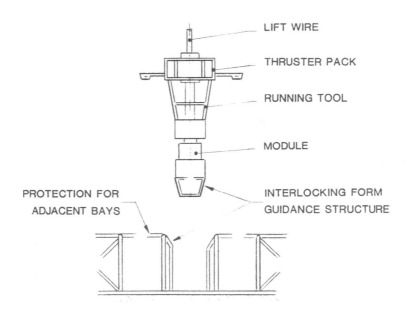

Fig. 5   Elements of a deepwater guidelineless running system

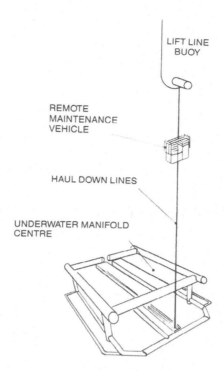

LIFT LINE
BUOY

REMOTE
MAINTENANCE
VEHICLE

HAUL DOWN LINES

UNDERWATER MANIFOLD
CENTRE

Fig. 6 Hauldown system as used on the Central Cormorant UMC remote
maintenance vehicle

stab prior to pulling the equipment into position, as shown in Fig. 7.
The body of the equipment itself can be used for guidance, as shown
in Fig. 8. Alternately, protective frameworks can be used for docking
with the equipment being slid across on a carriage, as shown in Fig.
9. These methods are best suited to neutrally buoyant packages with
their ability for fine positioning. The methods shown in Figs 7, 8 and
9 have been developed by FUEL for ESSO's Deepwater Integrated
Production System (EDIPS).

Whatever the guidance system, it must have a degree of compliance
to compensate for misalignment, equipment tolerances, etc. This can
be provided in the guidance/docking system, the tool mountings or
the tool body itself.

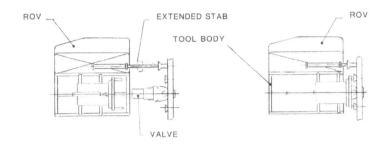

Fig. 7   Horizontal positioning using an extended stab and pull-in

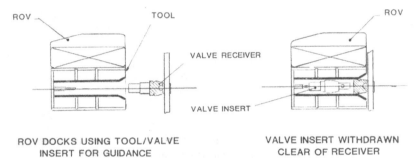

Fig. 8   Horizontal positioning using the equipment and tool body
for guidance

Tooling

Intervention tooling is an area which requires significant development.

Limited tooling exists for ROVs, but it is mainly for inspection, or
general-purpose manipulation. As conceptual designs for subsea
production systems emerge, specific tooling requirements are being
identified. These range from the relatively simple such as valve
override mechanisms, to complex tools of large size and weight for
removal of modules. All need development, construction and proving
before field use.

Non-ROV tooling carried by liftwire or drillstring, such as pull-in,
connection or module placement, also need significant development.

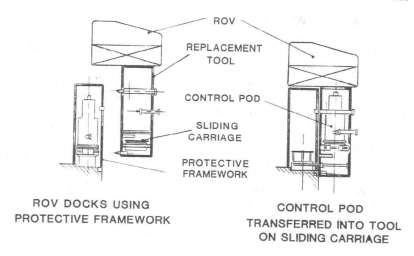

Fig. 9   Horizontal positioning using a protective framework for guidance

Back-up Systems

Although intervention systems will be designed to be as reliable as possible with in-built redundancy and overrides, alternative means of achieving some tasks, such as release of connectors and operation of valves is desirable and must be preplanned, if not actually implemented. These back-up methods will normally be of a different type to the prime systems, e.g. a mechanical override for a hydraulic connector. The ideal vehicle to operate these is likely to be the ROV.

IMPLICATIONS FOR SUBSEA PRODUCTION SYSTEM DESIGN

The selection of intervention methods of equipment will have a profound effect on subsea production system size, layout and operating philosophy. Template size is affected directly by the size of modules, tool packages and the ability to position them, as well as the provision of guidance and protection structures. Template layout will be affected by the need for access at certain points for observation or maintenance using tooling. Operating philosophy will emerge only when maintenance tasks have been identified and evaluated.

THE WAY FORWARD

The progression of deepwater intervention techniques is entirely in the hands of the major oil companies.

Gradual improvements in existing technology – ROVs, manipulators, underwater visual systems and inspection equipment – are bound to take place, but there is little money around for radical new development from small design houses, suppliers or operating companies.

There are many conceptual designs for deepwater production systems being pursued by the oil majors, but these are widely different in overall approach, and none will emerge as a significantly better solution for many years. In the meantime, however, it should be possible to identify various common intervention tasks and develop the tooling for these to the general benefit of the industry.

8

Production System Design: The Relationships Between Reliability, Redundancy and Maintenance Philosophy

I. J. A. Jardine, Baker Jardine & Associates Ltd

INTRODUCTION

There is an increasing awareness in the oil industry, particularly with respect to subsea engineering, concerning the importance of system reliability and maintenance assessment. In new field development studies the determination of a system's production efficiency forms an important part of the decision making process regarding a field's commerciality. In line with new developments in production concepts, methods and equipment, there have also been new approaches to the analysis of future field developments, integrating more closely the system design and economic appraisals. The recent slump in oil prices which has set back numerous field development studies will result in an even greater demand for in-depth assessment of production system behaviour. Attention will focus on improving both the design efficiency of a system and improving the confidence of the economic predictions.

This chapter provides an overview of the role of reliability, redundancy and maintenance assessment and how it impacts upon

Advances in Underwater Technology, Ocean Science and Offshore Engineering, Volume 10: Modular Subsea Production Systems
© Society for Underwater Technology (Graham & Trotman, 1987)

the design efficiency of subsea developments; a section is devoted to deepwater aspects. The underlying objectives are:

- to highlight the inter-relationship between reliability, redundancy and maintenance and their impact on system behaviour
- to demonstrate the role of new performance analysis methods in assessing and optimizing production systems via practical examples

PERFORMANCE ANALYSIS

Performance analysis involves the assessment of the overall behaviour of a system considering its design, maintenance and any external parameters which affect the functionality of the system over its working life. Reliability, maintainability, availability, operability and productivity analyses need to be integrated within a system performance study to assess the system's life cycle characteristics. Combining performance analysis with a costing exercise permits cost-effective system design with the potential of optimizing design efficiency.

Some examples are given within this chapter based upon data generated using computerized performance simulation techniques. These performance simulations have been undertaken using the MAROS computer model. MAROS is one of a new generation of computer packages developed to assist in system design. Performance simulators, such as MAROS, function by creating realistic life-cycle scenarios of proposed production systems.

A life-cycle scenario is a collection of events occurring during the anticipated life of the system, which conform with the system's logic and reflect its particular characteristics. The events themselves can cover a wide range from the normal (expected), such as; failure and repair of equipment, routine inspection and well workover, etc., to the abnormal (undesirable), such as; loss of well or loss of pipeline, etc.

The MAROS model can cope with scheduled, unscheduled and conditional events. Scheduled events occur under direct control of the user and are typical of planned maintenance activities where the system (or parts of it) are shutdown for overhaul. The user specifies externally the time(s) when such events will occur. Unscheduled events are generated by random sampling from mathematical distributions. Equipment failure and repair are typical unscheduled events. Conditional Event Logic permits assessment of group effects and allows the user to define a change in system configuration or operability dependent upon specific group occurrences.

MAINTENANCE CONCEPTS

Remote Maintenance of Subsea Equipment

In this context, remote maintenance includes all subsea work, including inspection, which cannot be conducted or controlled from the production facility (fixed platform or floater). Remote maintenance must be carried out by a separate vessel (or vessels). The characteristics of such vessels have a very important impact on the productivity and operating expenditure (OPEX) of a subsea development: in particular, the capabilities of the vessel (the range of repair work which it can cope with) and its availability (the amount of time which it spends in-field). To improve overall efficiency the vessel characteristics must be tuned to the number of wells, their layout and ancillary equipment. To demonstrate the general trends which arise and to give some quantitative feel of the effects of tuning, a maintenance philosophy analysis was conducted on a modular wellhead arrangement as shown in Fig. 1. Two main parameters were varied:

- the number of wellheads being served
- the availability of the maintenance vessel per annum

The underlying assumptions for this analysis were:

- the vessel could carry out all necessary repair work, including downhole activities, i.e. the vessel had heavy lift capabilities
- no account was taken for planned well workovers
- all wells were producers of identical flow potential
- the hire period was a single duration, i.e. not split over several return periods
- the five well clusters were on separate work zones

Figure 2 illustrates the results of this exercise, showing information relating to the wellstream availability, on entry to a manifold, and also the effective use of the maintenance vessel expressed as a percentage of its available time in-field.

On inspection of the curves of Fig. 2, it is immediately obvious how sensitive the availability of a subsea production layout is in respect of its associated maintenance service. As may be expected, providing a dedicated vessel (i.e. available all year round) maximizes the

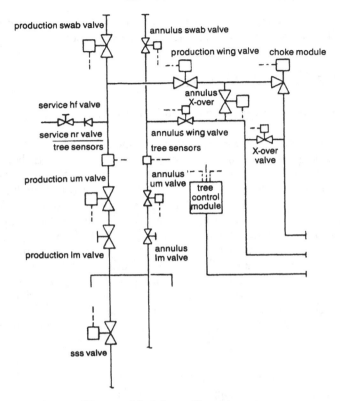

Fig. 1 Modular wellhead layout

availability of a subsea layout; however, in this example the utilization of the MSV is well below its optimum, i.e. there is still capacity to maintain more wells (equipment) without degrading availability unacceptably. Another five-well cluster could be maintained in this instance.

Further analyses were conducted to compare an *ad-hoc* maintenance philosophy for the five- and ten-well cases. It was assumed that a maintenance vessel could be chartered to carry out maintenance as required, and the vessels would take between two to four weeks to mobilize to the field. Results for the five-well cluster predicted a 92.5% wellstream availability, which would be a better solution than block chartering and probably more cost-effective than providing a dedicated vessel. However, for the ten-well case, an *ad-hoc* philosophy returned an 89% availability, which is relatively poor. Also, the charter costs were approaching the cost of a

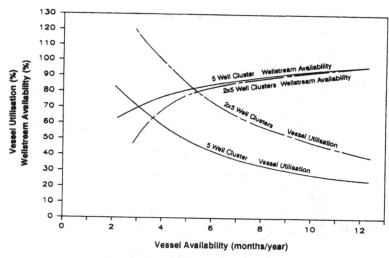

Fig. 2   Impact of vessel availability

dedicated vessel, thus indicating the need for a better strategy.

The results shown in Fig. 2 only convey average wellstream availability data, and provide no insight into short-term fluctuations. Knowledge of short-term variations in productivity is extremely important, particularly in gas developments where contractual production obligations must be met. Performance simulation techniques can provide this information for discrete time steps (e.g. months, weeks, etc.).

Figures 3(a)-(d) contain typical graphical output from the MAROS model, showing the predicted productivity profiles for the 2 x 5 well cluster cases. Notice the increasing range in monthly fluctuations as the maintenance vessel availability decreases.

Centralized Maintenance of Subsea Equipment

In contrast to remote maintenance, a centralized maintenance philosophy implies that all subsea work can be carried out from a single vessel on a fixed location. This vessel would normally be a floating production facility (FPF) with heavy lift capabilities, i.e. a drilling/production FPF. The concept of centralized subsea maintenance is very limited in its application: all wellheads must be in close proximity such that the maintenance vessel does not need to be relocated (other than within its allowable envelope dictated by either its anchoring or riser systems). Such restrictions limit the areal

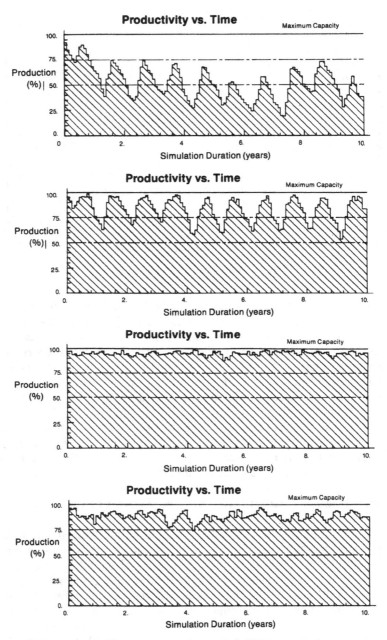

Fig. 3   (a) MSV 3 months/annum; (b) MSV 6 months/annum;
(c) MSV dedicated; (d) MSV *ad-hoc* charter

catchment of the wells and therefore this concept is reservoir dependent.

However, if permissible, this concept has attractions because in general a higher system productivity can be achieved compared with remotely maintained systems due to the "readiness" of the maintenance services. Production gains are achieved by taking advantage of the following:

- repairs can be tackled more quickly, resulting in reduced waiting time
- maintenance services are available all year round
- the large drilling/production FPFs offer less weather downtime than smaller remote maintenance vessels
- the centralized wellhead layout offers more scope for concurrent repair work
- there is less risk of external damage to subsea equipment, i.e. dragging anchors, trawl board damage

In addition to improved productivity, centralized maintenance concepts also offer lower OPEX because additional maintenance vessel hire costs have been eliminated.

Hybrid Maintenance Concepts

There can clearly be a combination of remote and centralized maintenance concepts. A drilling/production FPF may have a few satellite or remote well clusters which would necessitate a separate maintenance vessel. The ultimate choice is influenced, once again, by the reservoir features, and it is necessary to be able to analyse the performance of the various proposals.

MAINTENANCE STRATEGIES

A range of maintenance strategies can be applied to a system in an effort to improve its overall cost-efficiency. However, most of these are still very much at the theoretical stage as far as subsea technology is concerned. To develop a maintenance strategy for a given subsea development it is first of all necessary to be able to quantify the pro's and con's of alternative strategies at the conceptual development stage – hence the need for prediction methods.

While some recent subsea developments have operated "adequately" having no specific maintenance policy other than "fix it when it fails", it is likely that these developments would increase their profitability with a more sophisticated maintenance policy. The extent of improved profitability through better planned maintenance can vary considerably depending on the production system and reservoir characteristics. Indeed, it is possible to become counter productive through excessive maintenance.

Modularization of Equipment

The grouping of individual pieces of equipment into composite modules is a well known means of improving system maintenance. Modularization offers the following benefits:

- a reduction in the actual repair time of failed equipment via quick release modular connections and better access, etc.
- a reduction in the number of different subsea tasks to be carried out (useful when designing remote control vehicles)
- a reduction in the number of different types of spares required for system upkeep.

Preventative Maintenance

Subsea preventative maintenance involves the renewal of equipment on a planned basis prior to the equipment's failure. Such preventative maintenance would normally be associated with intervening a well at regular intervals. Advantage would be taken of summer weather windows, platform shutdowns, etc. when forming a preventative maintenance schedule. However, the frequency of the preventative maintenance and the extent of equipment replaced requires investigation.

Opportunity Maintenance

Opportunity maintenance overlaps preventative maintenance, yet there is a clear distinction between these forms of maintenance, as follows:

"Opportunity maintenance takes advantage of a situation in which certain services are available and certain criteria prevail to merit 'opportune' replacement of equipment prior to failure."

Only certain equipment should be considered for opportune replacement, and it is most successful on equipment which requires special repair services or equipment retrieved to the surface to permit access to other maintenance work.

In the next section, a quantitative example is given for a flexible production riser system incorporating redundancy and maintenance strategies.

The Role of Redundancy

Introducing redundancy into a production system is a means of offsetting the effect of specific failures on system availability. Redundancy is achieved by building over-capacity into a system via parallel training of equipment, and such equipment may be in either an active or passive mode. Providing redundancy is intended to improve system availability and hence productivity. However, redundancy also reduces system reliability and imposes increased requirements on system maintenance. It is possible therefore for redundancy to be counterproductive. Counterproductive redundancy is more likely to occur in subsea applications than others because of the nature of some subsea maintenance, i.e. the long waiting times before repairs are carried out. The foregoing may be difficult to digest; hence consider the following example of dual-choke modules on satellite wells.

Twin chokes are introduced to the wellhead to maintain control of the well when one choke fails. The reliability of the satellite wells is reduced by introducing more equipment, particularly the choke modules themselves which have relatively low reliability, and there is also additional valving required to permit the switching of the flow through a specific choke. Consider a situation in which the subsea maintenance capability in the field is restricted to several weeks per annum. There will be more maintenance work to be carried out on the dual-choke wells than single-choke wells because of the overall lowering of wellhead reliability; hence the maintenance vessel may spend more of its valuable time maintaining back-up chokes at the expense of equipment causing direct production losses.

This hypothetical situation is quoted as a caution to introducing redundancy liberally: it is not intended to generalize that redundancy subsea is worthless, nor dual-choke solutions for that matter. It serves to demonstrate the interrelationship between reliability, redundancy, availability and maintainability.

Consider a flexible production riser system as shown in Fig. 4. The riser bundle performs five functions:

- production
- export
- water injection
- test
- controls

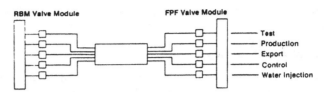

Figure 4a : Single Bundle with No Redundancy

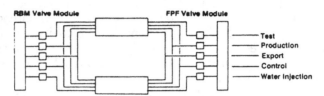

**Figure 4b : Twin Bundle with 100% Redundancy
(Export, Production and Controls)**

Fig. 4   Flexible production riser layout

The performance of the riser was simulated to investigate the following:

- The effect of providing redundancy for the production, export and control functions (see Fig. 4(b)).
- The effect of varying the riser maintenance philosophy.
  (a) Maintenance from the FPF itself (all year round).
  (b) Maintenance by specialist vessel on *ad hoc* charter assuming a 4-8 week mobilization.
  (c) Preventative maintenance: replace production, export and control lines every two years during platform shutdown, unscheduled failures as per (b) above.

Results of this investigation are given in Table 1. The table contains the predicted availability of the riser as a stand-alone system. It is evident from the results that, in this case, redundancy can offset very significant losses arising from lack of maintenance services and vice versa. In this example the line replacement policy improves riser availability. The final decision on which option to adopt requires a thorough cost-benefit assessment.

TABLE 1
Availability data for flexible riser system

Layout	FPF maintenance	Ad hoc charter	Planned strategy
No redundancy	98.5%	88%	90.5%
Twin bundle 100% redundant	99%	98%	99%

RELIABILITY

The term reliability in its strict definition refers to the failure process of equipment and systems, including the manner and frequency of failures. However, "reliability" is used frequently to cover all facets of system performance i.e. maintenance, availability, productivity etc. Semantics aside, reliability or, rather, unreliability is the root cause of the majority of system production losses. If equipment did not fail, then there would be no concern over maintenance strategies, no need to build redundancy into systems, system operating costs would be reduced and productivity improved, etc. Hence there are considerable potential benefits to be realized by improving equipment reliability. But which equipment should be improved and by how much? Developing improved hardware reliability leads to higher capital cost of the equipment. If the increased cost of the more reliable equipment exceeds its benefits, then it is a waste of time, money and resources.

Given the considerable disparity in reliability of subsea equipment from low reliability items such as chokes to the highly reliable wellhead connectors, there is a tendency to consider equipment only at the low end of the scale.

When choosing which hardware to further develop and improve, it is necessary to look beyond the equipment's unreliability and consider its failure impact, its method and time to repair, i.e. the equipment's overall importance within its proposed system environment.

Performance analysis methods can be used to predict potential benefits of hardware improvements and also to define reliability targets prior to embarking on development programmes. Figure 5 illustrates the effect on system productivity of increasing the reliability of subsurface safety valves (SSSV). These results were extracted from a floating production study on a 16-well field development which investigated both remote maintenance (production-only FPF) and centralized maintenance (drilling production FPF) concepts.

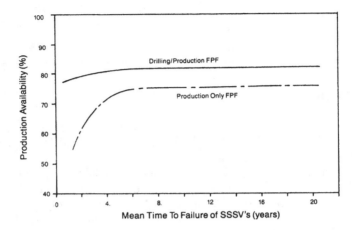

Fig. 5   Sensitivity analysis for SSSVs

Interpretation of the results in Fig. 5 conveys the following:

- Improving equipment reliability has different pay-offs depending on the system environment, i.e. the SSSV's in the drilling/ production FPF were not as "critical" as similar ones in the production-only FPF.
- For the production-only FPF, there is a critical reliability threshold for the SSSVs around 5-6 years mean time to failure (MTTF), i.e. the reliability of the SSSVs should be in excess of 6 years, but there are virtually no benefits in increasing the MTTF beyond 6 years.
- For the field investigated, the centralized maintenance concept is considerably more productive than the remote maintenance alternative.

Note that both concepts suffered 10% production downtime due to the tanker export system.

DEEPWATER ASPECTS

With increasing water depth and a more severe environment, subsea maintenance tasks become more difficult, more time-consuming, more costly and suffer from increased weather downtime. It is inevitable, therefore, that the availability of deepwater subsea developments will be degraded compared with their shallower water counterparts, while incurring higher maintenance costs (OPEX). Development effort is required to combat this trend. The parameters involved are the same for deep and shallow water; however, their scope and benefits vary as follows:

(1) *Modular packaging of equipment.* Careful consideration of equipment packaging may lead to reduced repair times and less frequent interventions. However, the scope for improvement is limited given that the modularization concept is already practised.

(2) *Vessel and tool design.* Improving the facilities and station-keeping of both surface and subsurface vessels will lead to reduced weather downtime. Consideration should be given to reducing or even eliminating the need for surface support vessels by further development of submarine technology.

(3) *Planned maintenance strategies.* Conducting operational research using performance analysis methods to establish the most cost effective maintenance policies for specific developments. The *ad hoc* approach will become less effective with increasing water depth.

(4) *Redundancy.* Providing equipment redundancy is always a possibility. However, in deep water, the advantages of redundancy will diminish and the philosophy of "simplicity" in design will prevail.

(5) *Reliability.* Probably the greatest benefits will be realized by improving equipment reliability, thus reducing the subsea maintenance burden. Gathering of equipment reliability and repair data must be continued to provide more extensive databases for use with performance analysis tools.

CONCLUSIONS

This chapter has illustrated the complex interplay between the parameters affecting the performance of a production system. The interrelationship between redundancy, reliability and maintainability has been demonstrated using typical subsea technology examples. To be able to convey this information it was necessary to have suitable analysis tools.

With new performance analysis tools available to assist system designers, emphasis can now be placed on attaining optimum design efficiency rather than just technical feasibility. Such optimization can be achieved through careful tuning of the numerous design parameters, taking advantage of the trade-offs between reliability, redundancy and maintainability. By investigating system behaviour more thoroughly, better predictions can be made regarding the productivity and OPEX of proposed developments.

The ramifications of the foregoing will lead to increased confidence attached to the overall project economics – a step in the right direction to encourage further developments (marginal, subsea or otherwise).

REFERENCES

Jardine, I. J. A. 1986. Computer simulation techniques applied to the design optimization of offshore field developments. *Petrolm Rev.*, February.
Kumamoto, H. *Reliability engineering and risk assessment.* Englewood Cliffs, NJ: Prentice-Hall.

9

New Technology for Modular Subsea Manifold

J.-P. Roblin, Alsthom-ACB

For deepwater field developments, subsea production systems are frequently envisaged by operators. A basic architecture consists of a central manifold gathering the production of several wells (satellite and/or template wells) and exporting it to a process facility. This scheme allows a large flexibility for the field development. Today, installed subsea manifolds are not very numerous. However, their number is expected to increase in the next couple of years.

The need to have cost-effective systems led oil companies to revise their philosophy: reliability, simplicity and maintainability are now the main objectives in addition to the reduction of investment and operation costs. For that purpose, new concepts and new technology for subsea production equipment have to be proposed. New ideas such as insert components, running tools or specialized ROVs, flowline/umbilical pulling-in and connection systems, operating robots, and new control system technology has been investigated and are presented in this chapter.

Advances in Underwater Technology, Ocean Science and Offshore Engineering, Volume 10:
Modular Subsea Production Systems
© Society for Underwater Technology (Graham & Trotman, 1987)

WHY MODULAR SUBSEA MANIFOLDS?

Drilling is now performed in deeper and deeper water, and any hydrocarbon strike in great water depth leads operators and equipment suppliers to develop various systems or concepts to bring such fields on stream. In particular, subsea production technology is more and more considered as a viable alternative when the water depth increases. It can permit the replacement of huge fixed platforms by simpler equipment lying on the sea bed. Moreover, subsea technology permits a better exploitation of shallow reservoirs: satellite wells are drilled vertically into different parts of the reservoir, which could not be reached by deviated drillings from a single fixed platform.

Such subsea wells have to be linked to a surface facility. Two alternatives are possible: either individual flowlines link each well independently to the surface, or production of several wells is brought to a central point where it is commingled in a common export line towards a surface facility.

Since it is normally required that it shall be possible to isolate each well from the others, the junction of the satellite well flowlines and the main export line will require valves. The structure receiving the valves, the piping, the flowline connecting equipment etc. is the manifold. It is often also a central point for other functions, such as TFL-servicing, dispatching of water injection or gas-lift lines to certain wells, and dispatching of remote control links to the wells. The manifold structure may also include subsea template wells.

In all cases, manifolds aim at reducing the number of production and control lines, and limiting (or suppressing in some cases) the surface equipment.

Subsea equipment needs maintenance. As it is considered difficult or even impossible to repair seabed equipment, a modular arrangement is used allowing the retrieval of defective equipment which can be repaired or replaced on the surface. That reduces the duration of sea operations and improves the ability to maintain equipment in good and safe conditions. Another reason for interest in a modular arrangement is the flexibility given for a field development. Additional producing wells or injection wells can be envisaged in a step-by-step development.

STATE OF THE ART

To date, installed subsea manifolds are not very numerous. However, their number should increase in the next few years. A distinction must be made between experimental manifolds and field-installed production manifolds.

Experimental manifolds are designed to prove the feasibility of specific technological choices (in particular, diverless installation and maintenance at great water depth). For instance, this was the case of the S.P.S. (Submerged Production System) installed by Exxon in the Gulf of Mexico; maintenance of components was accomplished by replacement by the MMS (Maintenance Manipulator System). It is also the case of the Skuld station, currently operated by Elf-Norge in Norway. It is a modular manifold controlled by means of an electrohydraulic multiplexed system from a platform located 20 km away. Handling is performed by a special tool, the COM (Connecting Operating Module), which also performs connections between modules.

Field-installed production manifolds are specifically designed for the exploitation of existing fields. They are often associated with a programme of tests of specific technological items. Most of them are wet manifolds (i.e. where all components are in contact with the surrounding seawater), as opposed to the dry concept where critical equipment is enclosed in a dry chamber where it can be maintained as if it were at the surface. Some of the existing field-installed manifolds are briefly presented hereafter as examples of what has been done to date.

Argyll is a wet diver-assisted manifold installed by Hamilton Brothers in 1975. It is located under a floating production facility (FPF). Garupa, operated by Petrobras off Brazil, is a dry manifold handling the production of seven satellite wells: commingled production is exported to a loading tower. Grondin, operated by SNEA (P) off Gabon, was a three-well diverless system maintained by means of a rail-mounted robot. Bonito, producing for Petrobras off Brazil in a 192 m water depth is the deepest manifold so far installed. It is located under a FPF. Buchan (operator BP), producing since 1981, is handling the production of seven wells which is exported to a FPF. The Cormorant UMC operated by Shell is a diverless nine-well system derived from Exxon's SPS: maintenance is performed by a ROV (Remote Operating Vehicle) lowered from a surface vessel (see Fig. 1).

It must be noted that other subsea manifolds are currently being considered or developed for fields such as Balmoral, Highlander, Troll, Oseberg, East Frigg and Tommeliten.

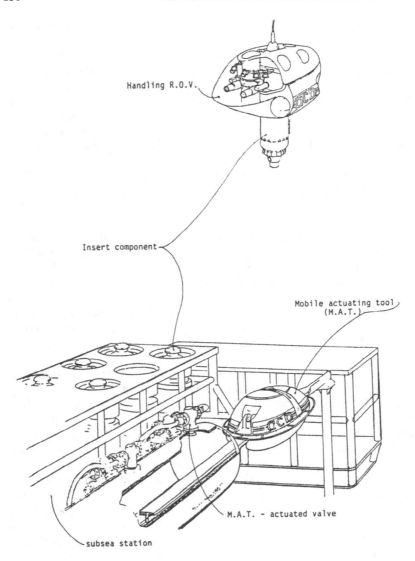

Fig. 1   Subsea station with insert components and MAT

MANIFOLD CONCEPTS: TRENDS AND EVOLUTION

Several manifold concepts have been studied or are currently being studied. It seems that oil companies are no longer concentrating their efforts on dry subsea completions. Mobil, however, is promoting dry subsea production systems with its Subsea Atmospheric System (SAS).

The opposite of the dry concept with human intervention at the bottom is the wet-modularized manifold concept with no diver intervention. Maintenance is performed by replacement of a module comprising several critical components such as valves, chokes or remote control equipment. This approach is somewhat different to that used by Exxon with the SPS, or Shell with the UMC, or Elf on Grondin experimental field, where a robot is lowered onto the manifold to change only one component. A typical wet-modularized manifold was the MSPS (Modular Subsea Production System) proposed by ACB in 1981: a completely guidelineless and diverless manifold handled the production of up to eight satellite TFL-serviced wells, it was either a riser base manifold or a remote manifold. The philosophy of CFP-Total, in association with IFP and Statoil, in the Poseidon project, started in 1982, goes even further: production is exported to the coast (150 km away) with the help of diphasic pumps and no surface installation is required. Another project has been developed by Shell, with its DIMOS (Diverless Installable and Maintainable Oil Production System), a modular manifold, in which modules are installed by means of a ballastable vehicle and where an acoustic remote control system is envisaged.

From 1976 to 1981, ACB was deeply involved in a programme carried out by French oil companies, the TEPMP Programme (Techniques et Equipements de Production en Mer Profonde). TEPMP phase 1 was completed with successful sea trials of the installation of a flowline between a satellite well and a manifold of 250 m of water. In 1983, ACB took over this programme and started a second phase. Advised by oil companies (Tenneco, Sohio, Marathon, Gulf and BP), ACB designed a modular subsea manifold and looked for new original solutions. This TEPMP phase 2, carried out by ACB, is presented in the present chapter.

TEPMP SUBSEA MANIFOLD

The outline for the study of TEPMP phase 2 consists of a hydrocarbon field in 600 m of water, located 3 km from a surface facility to be exploited by 10 subsea TFL-serviced wells. Provision must be made for converting any of these wells into injection or gas-lift activated wells as production declines. Due to the depth, no divers shall be required for subsea equipment installation and maintenance.

A possible field development outline is as follows: five wells are drilled from a remote subsea template and five satellite wells are individually connected to the template. Production is commingled and exported to the surface facility. Spare slots on the manifold permit the installation of an additional template well and the connection of an additional satellite well flowline.

All of the study was carried out keeping in mind the following general philosophy:

- a search for reliability through simplicity
- the prevention of pollution
- research into security for the whole system
- a search for a layout leading to the simplification of maintenance and installation operations from the intervention vessel
- wet architecture
- possible retrievability to the surface for maintenance of all active equipment subject to failure: no repairs to be performed on the sea bed

The study comprised two parts. First, a basic version of the manifold was designed according to the specification. It took into consideration the following specific philosophy:

- The use of field-proven equipment and technology.
- Procedures derived from drilling technology. In particular, vertical connections are considered. Modules including active equipment are installed from a DP drillship by means of a drillstring.
- As compact as possible an arrangement of the equipment: in particular, multibore connectors and a forged block integrating valve are used.

These philosophical choices led to a manifold using state-of-the-art components and technology. Alternatives to this basic version were studied according to the same philosophy in order to show the incidence of some specifications such as TFL-service and the complete versatility of each well.

In the second part of the study, new technologies were considered. New directions were investigated, in particular:

- simplification of installation and maintenance procedures
- simplification of the remote control system, in order to make it simpler and more reliable

These investigations led to the design of new special equipment, such as insert components, mobile actuating tools, etc.

The results of these studies of different manifold architectures are presented in the following sections.

MANIFOLD ARCHITECTURE USING EXISTING COMPONENTS

Method

The general philosophy defined for the study of the manifold was applied to the following points:

- definition of a flow diagram compatible with the specification in order to identify the necessary oil equipment (valves, chokes and piping)
- the arrangement of this equipment in different functional modules (modularization)
- organization of a control system circuit compatible with the identified modules
- design of the modules and other equipment of the manifold
- provision of adequate procedures and installation equipment

Architecture of the Manifold Basic Version

The TEPMP manifold consists of the superposition of the following structures, made up of tubular elements:

- a base structure, anchored by piles
- a semi-permanent manifold (SPM) locked up on the base structure, integrating petroleum links between modules
- retrievable modules locked up in recesses on the base structure or the SPM

The modules include all critical components. Following modules have been identified: connecting modules (9 + 1 spare place), valve modules (11 + 2 spare places), choke submodules (10), diverter modules (2), template Christmas trees (5 + 1 spare place), control modules (2).

The architecture of the modules permits their diverless and guidelineless installation from a DP drillship (see typical valve module and its associated choke submodule in Fig. 2). Flowline installation and connection are performed with the procedures which were field-proven in May 1981 during TEPMP phase 1. A wholly retrievable electrohydraulic multiplexed remote control system permits the actuation of all active equipment from the surface facility.

The overall dimensions of the manifold are $L = 42.500$ m, $W = 28.250$ m and $H = 13.800$ m. Its total weight is 2250 T (modules included).

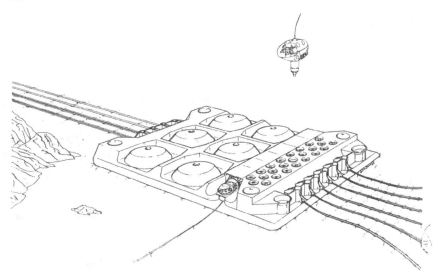

Fig. 2   Low-profile subsea manifold

A study showed that a limitation of the versatility of the manifold implied no modification of the architecture or the size of the manifold. Simplification of the components and of the piping does not justify a limitation of the versatility of the manifold.

A non-TFL version of the manifold was studied, with other technological choices similar to those of the basic version. It led to a smaller and lighter manifold ($L = 36.700$ m, $W = 21.600$ m, $H = 10.500$ m, weight $= 1570$ T). The difference, however, is not so significant and the choice of TFL or wireline service is one of philosophy rather than that of manifold equipment.

The TEPMP manifold, as described in this basic version, is a manifold using state-of-the-art components and methods. No technological aspects appear to require a completely new development. However, a weak point is the need for a fairly complex remote control system, due in particular to the large number of functions to be ensured by the manifold. Another weak point, inherent in the philosophy chosen for the study of this basic version, is the need for a fairly sophisticated maintenance vessel, which can moreover operate only within a good weather window.

In further studies of the TEPMP programme, solutions were searched for in order to limit these two weak points. The following directions were investigated, leading to new future architectures:

- simplification of the remote control system by reducing the number of functions to be ensured
- maintenance simplification by designing small modules that are easily installed or retrieved from a light service vessel
- installation simplification, by designing a small unitized manifold, using previous results
- drastic reduction of overall dimensions by use of insert template trees in a unitized manifold, using previous results

PROPOSED NEW TECHNOLOGIES

Two main new technologies were identified as particularly interesting for subsea manifolds:

- the "insert technology", leading to a simplification of installation and maintenance procedures
- the "MAT technology" (MAT = "Mobile Actuating Tool"), permitting a great simplification of the remote control system

These two technologies are complementary and their combination leads to a significant reduction of the size of the manifold. A new technology is therefore proposed for the flowline and umbilical connecting equipment so as to make it coherent with these two technologies.

Insert Technology

In a modular manifold, active equipment is located in a module: if it fails, the module is retrieved to the surface for maintenance of the defective equipment. The smaller the module, the easier and cheaper this operation is. The insert technology was conceived to make it possible to carry out maintenance operations by means of an ROV-type handling tool operated from a simple DP service vessel. The insert technology is the result of an ultimate modularization where each component becomes a module, retrievable independently from other modules from a light surface vessel. This permits a great simplification of the installation and maintenance procedures and a drastic reduction of the cost of these operations.
Insert components basically have the following characteristics:

- A vertical cylindrical external shell, guiding and positioning the insert component in a recess which is part of the subsea structure and which protects the active part of the component.

- A conical interface with the recess, through which hydraulic remote control and hydrocarbon flow connections are performed. Static metal-to-metal seals ensure good connections.
- An anchoring device, with hydraulically actuated dogs. For size reduction of this anchoring device, these dogs expand radially outwards in an inversed hub which is part of the piping integrated in the subsea structure.
- A handling head permitting handling by an ROV and through which hydraulic connections for the anchoring device actuation are performed.
- The active part of the component derived, if possible, from that of standard existing components.
- Syntactic foam, where possible, in order to lighten the weight of the component in water.

The insert components can be placed in retrievable piping modules or directly in the subsea structure.

Gate valves, choke valves, ball valves, TFL or pig diverters and electrohydraulic pods can cope with the insert technology.

Note that insert component technology has already been developed, essentially by Exxon and Shell. Valve suppliers, such as McEvoy, FMC and Vetco, have performed studies and some of them have manufactured equipment. Insert valves are installed on the Cormorant UMC.

However, there is a great difference between the insert components developed for the UMC and the insert components to be developed within the herein proposed project: Exxon and Shell adopted the insert technology so that valves and hydraulic modules can be horizontally retrieved and replaced by means of a sophisticated rail-mounted robot working on the subsea structure when needed. Alsthom-ACB proposes insert components whose arrangement comes from the minimum module concept. These insert components are vertically retrievable by means of a dedicated ROV, which permits much cheaper maintenance procedures (in particular, this ROV is operated from a simple service vessel).

MAT Technology

The most delicate equipment in a subsea manifold is certainly the remote control system. Petroleum equipment such as valves are normally guaranteed for a long lifetime: problems generally occur with the actuation system. In a modular version, if a problem occurs with the remote control system, a whole module must be retrieved to the surface, making it necessary to break the petroleum links. It is therefore interesting to dissociate the functional part of a component from its actuation part: the functional part of the valve can be a simple manual-type valve and a simple mobile actuating tool (MAT) can replace the actuators of the manifold valves.

This concept is particularly interesting: depending on the producing fluid, the estimated lifetime of manual-type valves can be long enough to make it unnecessary to locate them in retrievable modules, which permits a drastic reduction of the dimensions of the manifold (and therefore of the costs); moreover, the remote control system controls only the MAT instead of controlling as many hydraulic actuators as valves and is, by the way, greatly simplified.

This concept supposes a good location of the valves versus the MAT, i.e. an adaptation of the manifold to this concept. Furthermore, valves of the manifold cannot be fail-safe type valves: this is not a major disadvantage since manifold valves are production equipment, not safety equipment (for the same reason, the time necessary to actuate them is not a predominant factor).

Reduction of the Size of the Manifold

The reduction of the manifold size can be obtained by combination of the two technologies proposed above: critical equipment subject to failure or wear such as choke valves, TFL-diverters and electrohydraulic pods could be insert components, and gate valves or ball valves could be manual-type valves actuated by a MAT.

A complementary way to obtain this size reduction and to be coherent with the proposed technologies is the development of a small tool permitting the flowline pulling-in and connection to the seabed structure. A subsea friction winch developed by Alsthom-ACB is used for this purpose. Another method which has been envisaged is an adaptation of the satellite insert-Christmas tree technology to template Christmas trees. This would give the template/manifold a very low profile.

CONCLUSION

The proposed technology aims at improving the reliability of subsea production systems as well as reducing the investment and operating costs. Some described items are now being developed or have been proposed for different R&D projects or field developments. An insert gate valve will be manufactured and tested. Detailed designs for an insert ball valve and an insert TFL selector have been prepared and proposed to oil companies. A subsea friction winch was manufactured, tested and has already been used for offshore operations. New control pods are being studied by Alsthom-ACB and it is planned to modify them with insert technology. A development programme for the ROV designed for handling of insert components in under way. A large research effort is also being made to develop subsea tools, vehicles or robots which make it possible to perform subsea tasks in better and better conditions.

PART IV

10

Technical Development Requirements to Meet Future Needs?

C. J. Hedley, BP International Limited

INTRODUCTION

In trying to foretell what might be required for the future we face an almost impossible task, in that our perceptions are inevitably strongly influenced by today's prevailing conditions. One year ago "today" was $30 a barrel and in the UKCS deepwater production was just around the corner. "Today" is now $15 a barrel with deepwater production moving somewhat further into the future, and efforts tending to be focussed more on how to develop existing hydrocarbon reserves in shallower water at minimum cost.

The task is exacerbated by trying to decide where shallow water actually ends and deep water begins. For these reason this chapter will not be too constrained by the "deep water" label but will talk about subsea and modularization in a broader sense. It will discuss some of the key issues and technologies that are playing an important role today and are seen as requiring further development and improvement for tomorrow.

The chapter attempts to be wide-ranging and to give indications as

Advances in Underwater Technology, Ocean Science and Offshore Engineering, Volume 10: Modular Subsea Production Systems
© Society for Underwater Technology (Graham & Trotman, 1987)

to the fields of activity where developments will be needed, rather than talking in detail about one or two specific examples of modular equipment; to a large extent this will have been covered by the preceding contributions. As it is a form of "crystal ball gazing" it will inevitably be generalist in nature, although particular examples will be given where possible. Nevertheless, these will be mainly for illustrative purposes and in many cases are in hand already. They should therefore not be used as a "shopping list".

In keeping with the theme of the conference, the chapter is structured such that the types of modules that might be used are discussed first, followed by a review of some of the general issues which influence the choices to be made. However, the framework is mostly a matter of convenience to give a vehicle for raising some of the issues and inevitably most of the discussion will be covered in the earlier sections. The later ones will be used as a "catch all" and an opportunity to re-emphasise some of the points which will have already been raised in earlier presentations.

MODULES

It is first useful to consider three broad categories:

- systems modules
- equipment modules
- retrievable components

(or, if preferred, maxi-, midi- and mini-modules). The dividing lines between these categories are somewhat grey and, indeed, equipment modules might contain retrievable components, and systems modules might contain equipment modules. Nevertheless, there are some distinctions between these groups which can be used in subsequent discussions.

Table 1 gives some suggested examples of what might appear in each category.

Systems Modules

In thinking about systems modules we are concerned primarily with how we might develop any given hydrocarbon prospect. In addition to the subsea orientated examples given in Table 1, this will also involve a consideration of the merits of platform systems (be they fixed, compliant or floating), well completions, export systems etc. These can be packaged together in a number of different combina-

TABLE 1
Suggested categorization for subsea orientated "modules"

Systems modules	Templates, manifolds, riser bases, risers, satellite wells, export pipelines, flowlines, subsea slugcatchers and separators, subsea pumping units, control systems
Equipment modules	Xmas trees, valve modules, piping modules, flowline connections, control pods, TFL diverters, pig launchers/receivers, BOPs, wireline lubricators and winches
Retrievable components	Valves, chokes, pressure and temperature sensors, flow meters, valve position indicators, pig detectors

tions to give various field development options.

The merits of each combination will then be assessed against several economic criteria which will involve consideration of capex, opex, expenditure phasing, availability etc., but also a judgement (often subjective) will be made as to the technical risk. Thus, while cost reduction might be the main objective of incorporating novel technology, this is often counterbalanced by an associated increase in technical risk. A key issue in considering systems modules is therefore one of confidence, a point which cannot be emphasized enough, with technical risk assessment being of paramount importance. In such circumstances the approach to development is often a step-by-step one, with the more significant advances usually being driven by a strong economic incentive – whether an enhanced economic "prize" or the inability to develop a field economically without innovation.

It is worthwhile mentioning that this trade-off applies not only to the systems modules as such, but also to the whole concept of subsea operations *per se*, where major drawbacks are perceived to be the high cost of intervention and the low availability (poor reliability).

Having emphasized the points of "cost reduction" and "confidence", we can now consider what is probably the most significant area for cost reduction and hence a potential technical development target. This is in minimizing and possibly eliminating the need for new offshore processing capacity and hence offshore platforms.

In the mature hydrocarbon areas, marginal fields can be developed using remote (e.g. subsea) production facilities tied back to existing platforms in other locations, thus making use of any spare processing capacity on these platforms and the existing export infrastructure. In frontier areas, apart from minimizing the number of platforms required to develop any given prospect, this could mean transporting live hydrocarbons direct to shore for subsequent processing.

In both cases a key technology is that of multiphase flow transportation. It is important that we understand and are able to predict the behaviour of the fluids (e.g. pressure drop, liquid hold-up, slug size) with reasonable accuracy or, alternatively, if our predictive methods have significant uncertainty, demonstrate that the system is safe and operable for a wide range of conditions.

This becomes particularly important when we start considering longer distances. The slug size tends to become larger as bigger diameter pipes are used to limit overall pressure drop; moreover, the longer transit time (distance) permits the slugs to grow to a mature state. Riser-induced slugging can also become very severe.

Corrosion control can also become more complex, as large pipes with low flow velocities may lead to water settling out in the bottom of the pipe, as well as having segregated gas and fluid streams. This may require special materials or coatings, or special chemical inhibitor treatments (with and without regular pigging).

Hydrate formation may cause blockages of the line and again steps need to be taken to prevent it. Methanol or glycol injection might commonly be used, but careful consideration of operating conditions (e.g. pressure and temperature) may permit injection to be avoided by use of heating, insulation or depressurization techniques instead.

Wax deposition and sand transportation are other important related "hydraulic" considerations and, while solutions generally exist for resolving the problems, some of these may be overly conservative and may involve an associated cost penalty.

It is clear, therefore, that the whole subject of multiphase flow and its associated problems is very much a "black art" with significant scope for improvement, although the technical development areas are mainly concerned with improving our understanding and thus our confidence. However, there are related pieces of hardware, some of which exist and some of which require further development. A typical example is a subsea slugcatcher which, as well as smoothing out the effects of slugging flow, can often be made to act as a first-stage separator. Current designs are large and can be expensive, and as we move to longer and longer distances are likely to become larger and more expensive (due to increasing slug size). It would appear that there is scope for cost reduction either by reducing any conservatism or by adopting different designs.

In the former case this might be achieved by looking in more detail at the effect of the slugs on the whole process system rather than just the slugcatcher itself. In the latter case, the possibilities are open to speculation but, it is suggested, need to take a radically different approach from current ones.

Nonetheless it should be remembered that, if the problem is perceived to be limited to handling riser induced slugging alone, other solutions exist. For example, riser gas lift will be employed on our S.E. Forties field where existing process equipment on the FA platform is able to handle both "normal" slugs and the effective increase in GOR due to the injected gas. Slugcatchers should therefore not be seen as the only option.

Looking further to the future, multiphase pumping offers additional potential benefit, either by reducing the size of (or eliminating the need for) subsea slugcatchers, or by enabling fluids to be transmitted over even longer distances. Such pumps might either be downhole, surface or subsea – or indeed any combination of the above, depending on specific conditions. A land-based multiphase pump has already been successfully developed in conjunction with Stothert and Pitt, and is now looking for in-field service to build up its track record. The next step is to consider the sort of system necessary for subsea (remotely operated) duty. As well as further developments in actual pumping technology there is a need for suitable drivers (be they electric or hydraulic) and ancillary equipment (e.g. high power connectors).

This leads us to two intriguing questions: first "how far can we economically transport multiphase fluids?" and secondly, and more provocatively, "what is the future role of an offshore platform?" Regrettably, the answers to those questions are as yet unknown, but the challenge remains.

If subsea multiphase pumps can be successfully developed, then the alternative approach of subsea separation and separately pumped liquid and gas streams may not be needed. Other applications for subsea separators can be envisaged, however, and one possibility might be at the base of an existing platform (i.e. developing the slugcatcher concept further to keep additions to the topsides weight to a minimum). Another possibility might be a self-contained test separator unit to obviate the need for a separate test line.

An alternative approach to a test separator unit is, however, the satisfactory development of a multiphase meter. While the immediate target is to achieve a solution with sufficient accuracy for testing purposes, the ultimate challenge is to try and achieve fiscal standards (or at least come close enough to hopefully permit some sort of reasonable compromise). If this can be achieved, then the need to add process equipment to existing platforms solely to meter differing hydrocarbon streams for ring fence purposes can be eliminated.

In considering remote production of any kind, a key technology is clearly that of control. A large cost elimination offshore is the human one and the use of semi-automatic systems with minimal operator intervention must give cost savings; indeed, we might eventually be able to move to fully autonomous operations (at least for certain functions). Such systems may be particularly applicable to the pumping and separation scenarios just described.

As we move to longer and longer distances for remote production, the costs of the umbilicals will become an even greater element of the overall cost. Development of self-contained units with independent power sources and alternative telemetry techniques (possibly acoustic?) could also give significant cost savings. Acoustic telemetry could also give potential benefits over shorter distances and even on-template – primarily from the elimination of connectors.

Again, although it can be said that much of this technology already exists, it must be emphasized that *confidence* in the technology is the prime consideration. This inevitably means extensive testing, often at significant expense.

Equipment Modules

While systems modules do not have to be comprised of sub-modules, the opposite will generally hold in that equipment modules would form part of a systems module (for example, a Xmas tree would form part of a satellite well or template module). Confidence in these modules is still a key issue, but testing becomes somewhat easier to implement.

Some of these modules can be considered basic (e.g. Xmas trees, valve modules, control pods) and would be used in building up (say) production templates and manifolds. The challenge here is how to reduce both the capital and operating costs by simplifying the design, improving the reliability and making intervention easier. (Reliability and intervention will be discussed in more detail later.) Other modules might be considered specialist (e.g. TFL-diverters, pig launchers/receivers) and, while not ignoring the need to keep costs down, the challenge is more one of proving technical feasibility and establishing operability (the issue of confidence again). Unfortunately, the need for these "specialist" modules is not easy to forecast as the picture only unfurls as we investigate specific field developments in more detail.

Interconnection of these modules is an important consideration with the development of suitable, reliable multipath connectors being a vital activity. These connectors may be either horizontal or vertical and may interconnect various combinations of process, electrical and hydraulic paths. They need to be as compact as possible while permitting rapid connection/disconnection, integrity testing of seals, seal replacement *in situ* (ideally) and yet be able to tolerate relatively large misalignments. There is therefore always scope for what might best be called "cunning engineering".

In the categorization followed so far, the modules to be connected also include pipelines, and the development of flowline connection modules (be they for satellite well tie-ins or main export lines) is an area which holds scope for improvement. The type of connection system will obviously depend on the type of flowline and its installation method, but there is an increasing use/consideration of multiple lines (either bundled flexibles, towed steel bundles, or indeed reel-laid spiral steel bundles). Connection systems should therefore cover multiple as well as single bore cases.

As a more general issue, some degree of standardization of approach to packaging is desirable: this allows systems modules to be more easily constructed from equipment modules and may even permit the re-use of the modules in other applications. However, the standardization is more likely to be intra-company for the foreseeable future.

Finally, the development of any equipment module cannot proceed in isolation, it needs to be set in the context of a system, and all the associated hardware needs either to be already available or developed in parallel. For example, in our own DISPS programme (which has already been discussed) we are attempting to achieve all of the above objectives by going about the work in a systematic manner.

Retrievable Components

Retrievable components can be used as part of, or in conjunction with, or as a substitute for, equipment modules. Typically they have integral connection systems and are generally smaller in size than equipment modules. Their advantages lie in being able to maintain only that item that has failed without disturbing the rest of the system, their flexibility of arrangement, and the possible use of cheaper maintenance vessels. Against this needs to be weighed the increase in complexity and possible associated decrease in reliability.

Perhaps more than for equipment modules, the development of these items needs to be undertaken in conjunction with the intervention method. There is little point in developing a piece of equipment first and then deciding how it is going to be maintained second. The interplay with intervention system (be it an existing ROV, a special dedicated maintenance vehicle, or a maintenance tool) is therefore critical.

Apart from the "basic" components such as valves, chokes and pressure/temperature sensors, there may be a need for more specialist components such as flow meters, level switches (for slugcatchers and separators), position indicators, pig detectors, etc.

GENERAL ISSUES

The preceding discussion has categorized modularization and set out the developments associated with the "pieces of kit" needed to construct a subsea system. However, there are three more general issues which dictate the modularization approach to be adopted, initially from a systems viewpoint, but subsequently in deciding the extent of modularization and the relative merits of equipment modules and retrievable components. These are reliability, installation and intervention/inspection. They will now be reviewed and any implications addressed.

Reliability

The reliabilities of the constituent components of any subsea system are the major consideration in deciding how we might modularize for maintenance, and will dictate the number, size and relative locations of the modules. These design decisions are normally achieved by some sort of reliability assessment which would take into account the likelihood of failure, the resultant availability of the system compared with desired value, the method and cost of repair (including the surface support system), the cost of modularization and back-ups, etc.

Improved analytical models are becoming available and they permit a number of comparisons and sensitivities to be performed. Nevertheless, the analyses are only as good as the input data and, unfortunately, data on subsea equipment is not particularly good. For this reason the results must be carefully interpreted, and "what if" approaches are often adopted. Alternatively, target reliabilities might be established and test programmes put in hand to ensure these targets are met.

In more general terms however, there appears to be scope for improvements in reliability, not only in terms of average life but also in variability of performance between the same generic types (e.g.

different makes of gate valves) and indeed amongst the same make of valve (i.e. probability distribution of reliability). Recent in-house testing of a number of 4 in valves has shown some surprising results with significant differences in performance. Design improvements have been identified and information passed back to the manufacturers. Quality control has also been seen to be a major consideration.

We have every reason to suspect that the results of this valve testing programme would be repeated if we were to test other components (e.g. chokes, sensors). One of the prime technical development requirements is therefore to improve reliability of subsea components, a simple statement but one which may be subsumed to the alternative approach to accepting things as they are, and evolving sometimes complex solutions to effect their repair.

Installation

The cost of installation (and indeed maintenance) is driven primarily by the cost of the surface support equipment. Clearly the relative costs of such equipment are going to have a major influence on the modularization choices we might make. Fig. 1 is a simple representation of some of the types of vessels that can be used and the work they can perform, i.e. the types of "modules" they might handle. The challenge is to extend the capabilities of the smaller (i.e. cheaper) vessels such that they can perform a greater range of tasks (i.e. move further into the hatched area in Fig. 1) and also stay on station longer.

This might involve improvements in vessel capability (and enhanced dynamic positioning is one example), improved handling systems for the larger modules in worse conditions, specialist remote guidance vehicles, or just a change in perceptions about what can and what cannot be done. Examples of the latter might be trying to develop new ways of installing template structures using drill rigs, or alternatively considering in more detail how template structures can be constructed from sub-elements (i.e. interlocking elements).

As an example, the use of such structural sub-elements might find an application in stable ice-covered waters where modules could be installed through ice holes. In such cases there is likely to be a physical limitation to the size of the hole and hence the module. (The extent of such an application is, however, somewhat limited and the use of these techniques is less attractive in more severe ice conditions.)

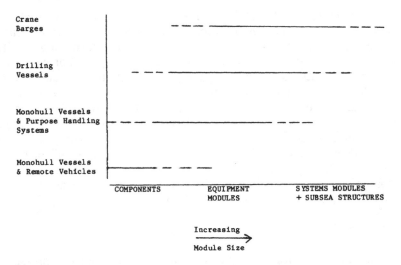

Fig. 1   Simple representation of differing types of surface support
equipment and the types of 'module' they might handle

Intervention and Inspection

The capabilities of intervention and inspection systems dictate the
layout of any subsea facility and will affect the design of many of the
components and the extent of modularization. Space must be left for
access for remote vehicles, be they swimming or tracked. Refinements to the capability of these vehicles may permit improvement to
the design of the subsea facility. These refinements might be either
in terms of being able to undertake a greater workload or being
smaller, more manoeuvrable and more controllable and hence
requiring less room.

Special purpose tools for ROVs are being developed, and will
continue to be developed, although as mentioned earlier this should
now happen in conjunction with component development such that
the two aspects are covered together.

The latest developments in robotics could offer some potential,
particularly in respect of manipulator design, but one suspects that
further experience of operating existing intervention systems is
needed before the need and benefits can be more clearly defined.

Image processing and image enhancement may also begin to
extend the capabilities of remote vehicles and, while it is early days
yet, may contribute to the eventual development of automatic or

semi-automatic navigation systems. Simulation models might also find an application, particularly if reliability does improve sufficiently to make intervention less frequent: operator training and re-familiarization would then need more consideration.

Finally, well service operations offer much potential for improvement, and current moves towards subsea wireline lubricators and wireline winches operated from diving support vessels should hopefully bring about cost reductions. The results of current and future field trials are awaited with interest.

SUMMARY

A number of issues have been discussed, some in more detail than others, but that is a function of today's perceptions. The main thrust must be towards cost reduction and, as such, the subjects of multiphase flow and remote control are suggested as being the most significant areas for technical development. The majority of the work is associated with confidence rather than new pieces of kit, but opportunities do exist for improvements in equipment.

Other significant areas for development include improvements in reliability, extension of surface vessel and ROV capabilities, development of improved connection systems, and the application of the latest robotic and IT techniques.

ACKNOWLEDGEMENT

Permission to publish this paper has been given by the British Petroleum Company plc.

11

Evolution of Esso Subsea Design

R. L. Hansen, Exxon Production Research Co.

INTRODUCTION

Industry-wide research and commercial installations over more than 25 years have advanced subsea completions into a technically mature option for offshore oil and gas development. Field experience now approaches 400 m water depths, with plans under way to go much deeper. As industry progresses into deeper water, a key requirement is the capability for diverless intervention for inspection, maintenance and repair of subsea equipment. Esso has been a leader in this area and has undertaken, since the 1960s, several major Research & Development programs as well as commercial projects to advance the state of subsea technology including diverless maintenance systems. The latest R&D effort, which draws on more than two decades of experience, is the Esso Deepwater Integrated Production System (EDIPS) development program undertaken in the UK by Esso Exploration and Production UK.

This chapter presents key design features of Esso subsea systems, which have made major contributions to the advancement of subsea technology. The evolution of Esso's subsea technology is

Advances in Underwater Technology, Ocean Science and Offshore Engineering, Volume 10: Modular Subsea Production Systems
© Society for Underwater Technology (Graham & Trotman, 1987)

described and the EDIPS program, as it relates to the subsea theme of this chapter, is also briefly discussed at the end, noting how the programme is pursuing several near-term design challenges.

EARLY GULF OF MEXICO SATELLITE COMPLETIONS

Exxon USA's first subsea well, designated 803 U-3, was completed in early 1964 beneath a platform in 18 m of water in Grand Isle Block 16 field. To simulate satellite well operation, this well was produced to and controlled from a platform one mile away. The initial diver-assist tree was later replaced with a more advanced design to prove operational capability without divers. The new tree featured: (1) remote flowline connection; (2) remote riser connection; (3) internal porting of all hydraulic control and electrical signal lines; (4) block tree construction utilizing a single casting for the entire tree body; (5) metal-to-metal seals throughout the production runs; (6) through flowline, or TFL servicing capability; and (7) electrical valve position indicators and pressure transducers. All of these features are used to varying degrees on the latest subsea completions today, which are conceptually very similar to this early completion. The 803 U-3 well operated successfully, including use of TFL for servicing, until it watered out.

Exxon USA's next underwater completion, designated 803 U-4, was a true satellite completion located in 18 m of water, again in Grand Isle Block 16 field, two miles from a platform. This project was industry's first completion of an underwater well with the rig off location. Prior to the offshore completion, a complete suite of TFL completion and workover tools was developed in an Exxon land test well. After the tree was installed and tested on the offshore well, the jack-up drilling rig was removed without completing the well. Then dual flowlines were installed, the well was perforated remotely using a TFL perforating gun, and the producing sand was plasticized using TFL techniques. Although some difficulties were experienced with the early design TFL locomotives, the completion was successfully brought on stream using TFL, with no rig intervention required. Many of the TFL improvements from this project benefitted Esso and others on later wells.

ESSO'S SUBMERGED PRODUCTION SYSTEMS (SPS)

Esso's major SPS effort was initiated in 1968 when a team was formed by Exxon USA to develop production capability beyond the reach of bottom-founded, surface-penetrating structures in the Santa Barbara Channel (SBC) off California. The resulting initial development concept was to produce subsea well clusters to platforms. Although the extension of platform water depth capability eventually negated the motivation for the SPS in SBC, the SPS technology development was continued and extended to meet deepwater Esso requirements anticipated elsewhere. Consequently, the SPS evolved into the system shown in Fig. 1.

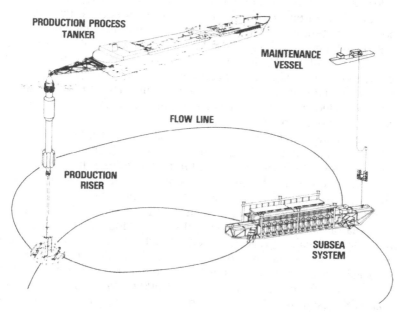

Fig. 1   Esso deepwater Submerged Production System (SPS)

The Esso SPS is a full capability, totally integrated deepwater production system. Each SPS template can be designed to accommodate up to 20 wells, or more if necessary. Drilling operations are performed using conventional deepwater drilling equipment and procedures. Production from all wells is combined on the seafloor template where it flows through flowlines, up a production riser and onto a processing and storage vessel which offloads to shuttle tankers. Alternatively, production can flow through pipelines to a platform or to shore. All production functions are controlled remotely from the surface facility using an electrohydraulic control and communication system. Wells are serviced using TFL tools pumped from the surface facility. SPS operational capabilities cover all production phases including (1) deepwater system installation, (2) development drilling, (3) well completion, (4) production and transportation, (5) well downhole maintenance, and (6) system maintenance.

Esso has chosen a surface-controlled, unmanned approach for underwater maintenance. The choice of unmanned maintenance in 1968 was a rather bold initiative because remotely operated underwater vehicles, commonly called ROVs today, were in their infancy. A maintenance manipulator is used to service hardware on the subsea template. Operationally, the unit is deployed from a workboat-type vessel to the SPS template where it is driven by remote control to the location of a service task. Maintenance is achieved by removal and replacement of specially designed valves or control modules. Similarly, a remotely controlled vehicle, launched and controlled from the surface, can perform service tasks on key components of the production riser.

Esso engineers considered the possibility of putting the SPS on a bottom-founded structure that would reach within diver access of the surface, then considered to be 180 m maximum depth. (This idea of using underwater platforms has recently appeared again by others who are re-examining the concept for deepwater development.) When Esso engineers were convinced that a successful unmanned manipulator could be developed, they concluded that it would be more economical to place the wells directly on a seafloor template and completely eliminate diver dependence. This proved to be an especially wise choice later on when Esso decided to extend the SPS design depth from the original target of 600 m down to 1500 m. The final SPS design is relatively insensitive to water depth over these depth ranges.

Before leaving the subject of maintenance, it should be noted that the development of remotely operated vehicles (ROVs) in recent years, especially those with significant work capability, is expected to greatly enhance diverless subsea system maintenance capability in the future. Some recent subsea designs by Esso and others have made extensive use of ROVs for installation and maintenance.

One interesting design feature of the original SPS is the optional capability to add a subsea separator and pump station, where needed for certain site specific conditions. For example, a pumping system with long pipelines to shore might compete favorably with a production riser and tanker for some conditions. The SPS separator/pump station was tested successfully on land, although a failed underwater electrical connector precluded subsea testing offshore. The failed electrical connector was subsequently redesigned and successfully tested.

There is an effort in industry today to develop subsea pumps, with the main motivation to pump full wellstream over long distances to shore. The recent effort by several groups to develop multiphase subsea pumps should improve upon the separator/pump arrangement originally developed by Esso. However, the power supply and electrical connection area will need careful attention to insure reliable operation. Also, economic solutions must be found for other technical problems associated with full wellstream transfer, such as the formation of solid hydrates at high pressures and low temperatures.

Exxon USA installed a three-well prototype of the SPS in 52 m of water off Louisiana in 1974 (Fig. 2). The purpose of this SPS pilot test was to evaluate systems hardware and deepwater operational procedures. Although the pilot test was in 52 m of water, the system was designed and fabricated for 600 m water depth service, and all installation, development and operational phases were conducted consistent with a full 600 m application.

As a result of $81 million ($200 million in today's dollars) and several hundred man-years expended over a 12-year period on the

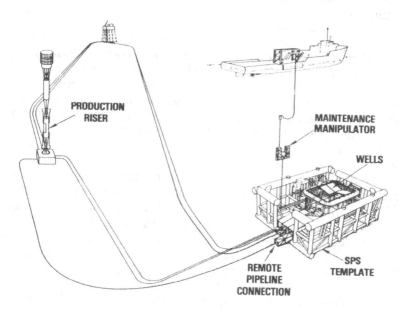

Fig. 2 Esso SPS offshore pilot test

SPS pilot test, Exxon USA concluded that deepwater procedures for installation, operation and maintenance had been demonstrated; that the feasibility of all of the equipment required for a full production system had been proven; and that the operating experience gained would serve as a basis for further extension of the technology. Esso SPS technology was ready for a commercial application.

SHELL/ESSO UNDERWATER MANIFOLD CENTER (UMC)

In late 1974 Shell and Esso formed a joint engineering and management team to develop an underwater production system for the hostile northern North Sea. The Shell/Esso UMC combines Shell experience from subsea satellite wells in Brunei and elsewhere with Esso SPS experience. Although the nine-well Shell/Esso UMC overall appearance is very similar to the three-well Esso SPS, many improvements and additional features were added to increase the reliability and flexibility of the system.

Figure 3 shows the layout of the UMC system as designed for the Central Cormorant field. The P-1 satellite well was brought onstream in January 1981, producing directly to the South Cormorant Platform. Successful field operation of P-1, which incorporates many key features of UMC wells, provided confidence for the subsequent installation of the UMC in 1982. The UMC itself is designed to support up to nine wells, each of which can be drilled directly through the UMC template, or drilled remotely as satellites which tie back to the UMC using flowlines and control umbilicals. This choice of well location feature, which was not provided on the Esso SPS, allows flexibility to accommodate reservoir size and shape uncertainties as development drilling proceeds. Another technology extension on the UMC is the use of insulated flowlines to control wax, hydrates, and emulsion problems that could result from produced fluids cooling toward seafloor temperatures.

Like the SPS, the UMC is designed for remote installation and maintenance. However, since the Central Cormorant water depth of 150 m permits ready diver access, both remote and diver assist maintenance capability was developed concurrently. Both have been used successfully for the minimal maintenance required so far. The remote maintenance tool is the Esso SPS maintenance manipulator, which was extensively refurbished by Shell/Esso for North Sea operations. When it was deployed on the UMC, it accomplished its assigned maintenance tasks expediently with no difficulty. Despite its successful operation, both on the SPS and UMC, the maintenance

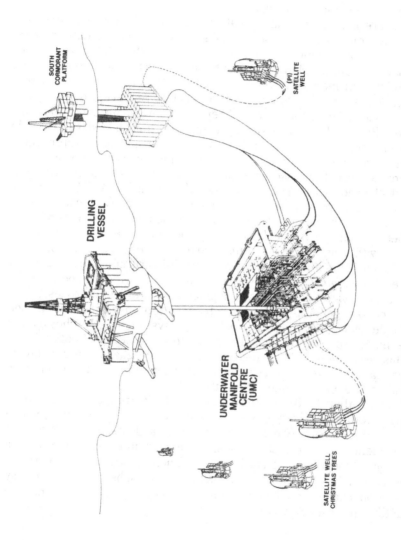

Fig. 3 Shell/Esso Underwater Manifold Center (UMC)

manipulator is large, heavy, and cumbersome to handle. Looking toward the future, Esso and others continue striving for improvement in maintenance techniques, particularly through the use of the smaller and more maneuverable ROVs.

There is an interesting debate among subsea engineers over the relative merits of component maintenance versus modular maintenance. Component maintenance refers to the ability to replace relatively small, failed components such as valves and control pods. Modular maintenance refers to relatively large, heavy modules containing many components which must be retrieved to the surface to replace a failed component. There is a common misconception that the SPS and UMC are strictly component-maintained systems. In reality, they are both hybrids, using both maintenance approaches. Although the SPS and UMC have many replaceable valves and control pods, they also have heavy modules, such as the hydraulic and electric power skids, and the Christmas trees. The SPS and UMC have proven both component and modular maintenance. Therefore, the designer has a choice, which will be governed by his specific functional requirements, his perception of industry experience with maintenance systems, and his business preferences.

Regardless of maintenance method, it is crucial to recognize that the design and maintenance of a subsea system, especially for diverless water depths, is an inseparable, interactive process. One can choose general-purpose maintenance tools available now, or one can develop special-purpose tools for his system, but the choice must be made early in the design process to avoid serious conflict later. This statement holds true even for simple shallow water diver-maintained systems, where diver participation in the design process will avert expensive difficulties offshore.

Two other UMC features, TFL and electrohydraulic controls, were both extensively developed and proven on the Esso SPS, and on early satellite wells by both Shell and Esso. Despite high confidence in these technologies, it is important to remember from a design standpoint that each system functional capability should be economically justified for each site-specific subsea design. Every function carries a cost and complexity that may or may not be justified for a specific site.

ZINC SUBSEA CLUSTER WELLS

A preliminary design was completed by Exxon USA for the subsea system shown by Figs 4 and 5. The plan was to install a five well slot

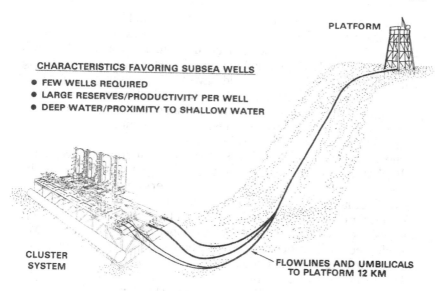

Fig. 4   Zinc subsea development plan

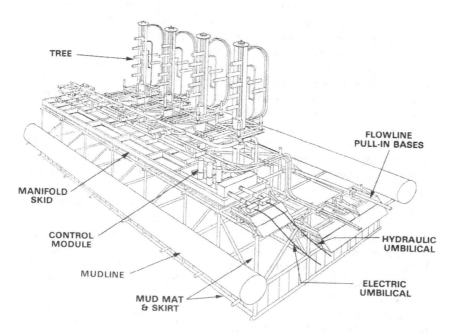

Fig. 5   Zinc subsea template system

subsea template to produce gas from the Zinc field in 460 m of water in the Mississippi Canyon area of the Gulf of Mexico to an existing platform 12 km away. Zinc is a good example of a field that will require a relatively small number of wells, where a surface piercing structure or floating vessel would have been much more expensive than a subsea system. Although the Zinc project has been put on hold pending gas sales and transportation arrangements, the design was carried to a sufficient level to illustrate the continued evolution of several key subsea design considerations.

The Zinc system shown looks very different from the SPS and UMC, but it incorporates many key technology items proven by the previous projects. Use of TFL for well servicing, and multiplex, electrohydraulic controls was justifiable for Zinc conditions. Due to the need for diverless installation and maintenance in 460 m of water, much of the diverless technology developed on the SPS and UMC was adopted for Zinc including flowline and umbilical connection tools, and tree running and connection equipment.

The main point of departure on maintenance was the choice of a commercially available ROV carrying specially developed tool packages, instead of the tracked maintenance manipulator used on the SPS and UMC. Maintenance tasks performed by the ROV would be the same as those performed by the SPS/UMC maintenance vehicle, namely, replacement of valves and control pods.

Choice of an ROV instead of a tracked vehicle for Zinc should not be taken to mean that an ROV will always be the best choice, although it seemed to ideally suit the Zinc requirements. More functional requirements, such as gas lift or reinjection, and water injection, would add more valves and require a more complicated manifold, which could possibly be more efficient to maintain with a tracked vehicle. To re-emphasize two crucial points made earlier, the system functional requirements must be firmly defined, and the maintenance system must be designed concurrently with the overall system layout.

The 460 m water depth at Zinc and the absence of commercial fishing or other potential interference permitted a system design with minimal protection. Since the trees are easily accessible, the preliminary design provides for replacement of tree valves using the ROV system. This eliminates the costly need to pull the tree to replace a failed valve, which is the maintenance method for all deepwater subsea trees installed so far. The replaceable valve configuration proposed for the Zinc trees is the same as that used on the SPS/UMC manifolds, and the removal and replacement methodology is well established.

One other Zinc maintenance provision not possible on the UMC is the ability to remove the manifold. Although the probability of needing to remove the manifold is small, this capability was achieved without adding much complexity to the system design. This task was simplified because the small number of wells permits a simple, straight manifold. In addition, the probability of needing to retrieve the manifold was minimized since most of the valves were located on the trees instead of on the manifold.

Flow analysis for Zinc flowlines showed that the production stream would cool down to the hydrate formation region before reaching the platform, even using state-of-the-art flowline insulation. This problem was addressed by providing continuous methanol injection, which was economically feasible at Zinc due to the very small volume of water produced with the gas. Finding improved methods to economically suppress hydrate formation for full wellstream transfer from subsea systems is a priority work area.

To overcome quality assurance (QA) problems experienced with hardware on the SPS and UMC projects, a formal QA program was established by Exxon USA for the Zinc Project. Ability to meet project QA requirements was a necessary condition for bid qualification. Reliability of subsea equipment must substantially exceed that of surface equipment due to the high cost of subsea maintenance. Quality assurance must be methodically incorporated into all facets of the design, manufacture, installation, and test of all components and subsystems.

ESSO AUSTRALIA SUBSEA WELLS

The Cobia-2 subsea completion was installed in the Bass Strait by the Esso Australia Ltd/BHP Petroleum Pty Ltd Joint Venture during 1978 and 1979 (Fig. 6). It produced nearly 2 million barrels of oil by natural flow from 1979 to 1983, when it was superseded by production from the Cobia-A platform and subsequently plugged and abandoned in 1984. OTC Paper 5315 summarizes the operational experience, including the post-mortem analysis.

From a design standpoint, Cobia-2 had many features of earlier subsea wells by Esso and others. The 76 m water depth permitted diver-assist installation and maintenance. Multiplex, electrohydraulic controls and TFL well servicing were provided. Operation of this well would not have been possible without TFL, which was frequently needed to clear the uninsulated production flowline of severe wax deposition. Over 20 wax-inhibiting chemicals were tried

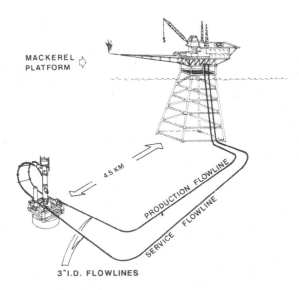

MACKEREL
PLATFORM

4.5 KM

PRODUCTION FLOWLINE

SERVICE FLOWLINE

3"I.D. FLOWLINES

Fig. 6   Cobia 2 subsea completion

unsuccessfully before reverting to a schedule of 12 hours production
followed by 5 hours of TFL wax scraping. Over 770 TFL scraper
runs were made over the 4 year producing life. The TFL system for
Cobia-2 was also used to lockout a leaking tubing retrievable
subsurface safety valve and install an insert safety valve. No re-entry
of Cobia-2 was required from a drilling vessel over the 4 year
producing life.

Esso Australia has completed preliminary designs for future
subsea completions in the Bass Strait, although project plans have
been deferred due to the adverse economic climate for oil and gas
development. Where justified, insulated flowlines have been designed to minimize wax deposition. In addition, chemical injection
points at the tree will permit injection of hydrate inhibitor, corrosion
inhibitor, and wax solvent. Some wells may require gas lift, and TFL
will be provided where justified to service flowlines and downhole
equipment.

Although the future Bass Strait subsea wells studied so far are
within diver depth, serious thought has been given toward the use of
ROV support for certain maintenance tasks. Advances in ROV
technology and the commercial availability of several competing ROV
systems may make diverless ROV maintenance economically com-
petitive with divers for many tasks, even in shallow water.

ESSO DEEPWATER INTEGRATED PRODUCTION SYSTEM (EDIPS)

The EDIPS development program is a major R&D initiative launched by Esso Exploration and Production UK in 1985 using UK contractors, with input and technical coordination by Exxon Production Research Company, Esso UK's research affiliate. EDIPS brings together several major subsea and surface production systems into an integrated system. The program objective is to assess, and extend as required, technology available to produce hydrocarbons in 600 and 1050 m water depths in harsh environments.

It is not the intent of EDIPS to develop a solution for site-specific application. Rather, Esso UK aims to develop potential building blocks of a total system and examine the integration requirements to bring together the building blocks into a system. The flexibility to configure one or more of the building blocks into a production system for site-specific application as needed is an important consideration.

EDIPS is a four-phase program with the first phase, completed in August 1985, defining the overall configuration shown in Fig. 7. It includes a TLP, two underwater manifold centers (UMC), two satellite wells, a single anchor leg mooring (SALM) with a production manifold at the base, and a floating production and storage unit (FPSU). TLP production is processed on the TLP and transported to the FPSU via the riser base manifold. Production from the subsea systems is commingled at the riser base manifold and produced to the FPSU for processing and storage.

Esso UK completed the second phase in June of this year, which included a conceptual design of the system and, more importantly, identified the critical components and systems requiring further work to improve confidence in Esso's readiness to use the concepts in commercial application. Not too surprisingly, one area where significant work is being pursued in Phase III is the development of diverless maintenance systems. While the systems being developed are required for deep water, a significant motivation to pursue the designs in the current business environment is their potential cost-effectiveness in shallower water depths, as previously noted in discussing Esso's subsea designs in Australia.

The major EDIPS contribution to subsea technology is the remotely operated, diverless-maintained manifold at the riser base, which permits manifolding of two UMCs and two satellite wells. In the development of the diverless maintenance systems, Esso is looking to draw and improve upon concepts developed from past work on other systems. However, the weathervaning FPSU tanker

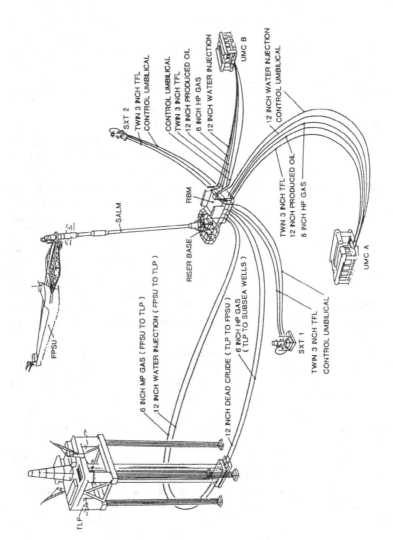

Fig. 7 Esso Deepwater Integrated Production System (EDIPS)

overhead, and the presence of the SALM offer some unique design challenges and opportunities. In order to minimize maintenance costs, the FPSU tanker is utilized as the deployment platform for both ROV and guideline deployed systems. ROV tool packages are being designed for replacement of smaller components (e.g. 3 inch valves and control pods) and guideline-deployed running tools for replacement of larger modules (e.g. control distribution modules) on the EDIPS riser base manifold. The EDIPS maintenance approach will thus take advantage of the best features of both component and modular maintenance.

Designs for the ROV deployed maintenance systems are to be developed further in Phase III, with the possibility of going into prototype fabrication and testing, if required, in Phase IV. Particular attention has been given to making the designs for the ROV-deployed maintenance systems for EDIPS simpler than those for the Zinc systems wherever possible. In addition to these diverless maintenance system designs, Phase III of the EDIPS development program will involve several other projects, all carefully selected to improve Esso's confidence in the technology in the near term. Several of the near-term design challenges, from Esso's viewpoint, are outlined below.

NEAR-TERM SUBSEA DESIGN CHALLENGES (5–10 YEARS AHEAD)

Although subsea technology is available today to permit deepwater development, there is considerable opportunity to increase confidence and reduce or better define costs through further engineering design and development, especially in the following areas:

- *Diverless installation and maintenance.* Although the SPS/UMC maintenance manipulator accomplished its objectives on both projects, there is incentive to reduce size, weight and complexity to simplify deployment and operation. Advances in industry ROV technology in recent years look promising here, and are being pursued further by Esso on the EDIPS project, and by others.
- *Marine production riser.* This is another high-cost subsystem for a remote, deepwater production system, such as EDIPS. Building upon design and field experience with Esso's SPS riser, and the Hondo and Fulmar Fields' SALM and OS&T systems, Esso is developing a deepwater production riser design for EDIPS. With maintenance as a major design factor, various alternatives such as

hoses and swivels were considered for high pressure fluid bypass of articulated joints. Flexible hoses have been selected for EDIPS. The maintenance systems for hose replacement will need development. Although not part of the EDIPS program, Esso is also evaluating dynamic positioning assist mooring alternatives which could lead to reduced costs.

- *Pipeline, flowline and control umbilical installation, connection and maintenance.* This area can represent a substantial portion of the total installed cost of a subsea system. Esso is working in this area both on EDIPS, and on other R&D projects by Esso Expro UK.

- *Hydrate control.* Most subsea systems must take hydrate control into account to one degree or another, and it can be a major design consideration for a large subsea system such as that used with EDIPS. Finding improved methods to economically control hydrate formation for full wellstream transfer is a priority work area.

- *Subsea equipment reliability and quality assurance.* Published SPS and UMC papers noted the need for further effort here by industry manufacturers. API has effort underway to develop standards for subsea systems, and to implement quality standards for oilfield equipment in general. Although oilfield equipment is generally reliable and designed to withstand abuse, subsea maintenance is sufficiently more costly than surface maintenance to warrant more effort in this area.

- *Design pressure.* Wellheads of most subsea systems are rated for 10 000 psi. Christmas trees of all but a few are rated for 5000 psi. Much design work will be needed on valves and connections to extend pressure rating for some future reservoirs. Esso is looking at pressures somewhat greater than 5000 psi for the EDIPS system.

- *Artificial lift.* Most subsea wells have been produced by natural flow, although some are being gas lifted. If operators anticipate artificial lift requirements for some future reservoirs, development work may be required, especially for methods other than gas lift. Very little information has been published on artificial lift of subsea wells.

- *Well servicing and workovers.* For non-TFL wells, most downhole servicing has been done using a fully equipped floating drilling rig. Cost reduction in this area would improve overall project economics. There have been some recent examples of successful servicing from smaller, lower-cost vessels, including one such system in the North Sea using a diving support vessel. Continued development of this technology would be welcomed by subsea well operators and designers. Major downhole workovers, where downhole equipment must be removed will still require a fully equipped floating drilling vessel. However, even here, careful planning and design of the completion and workover equipment can save time and reduce cost.

- *Surface controlled, subsurface safety valves (SCSSVs).* As mentioned above, wireline servicing from a floating vessel is expensive, and most wireline operations on subsea wells have been to service the SCSSVs. TFL provides a servicing method at lower maintenance cost than wireline, but TFL carries a capital cost penalty. Therefore, overall subsea well servicing costs could be reduced by improving the service life of SCSSVs. Operators and manufacturers need to work closely to choose the best valve designs for each application, and manufacturers need to continue improving the design and quality assurance of their products.

SUMMARY

This chapter has highlighted key design considerations of subsea production system technology by describing the evolution of Esso's subsea systems over more than two decades. Even though platform water-depth capability has been greatly extended in recent years, the use of subsea completions will continue to expand in response to several motivations: to augment platforms; to produce to existing infrastructure; to produce relatively small reserves to floating production systems; and to develop deepwater reserves. There is no universal subsea design. Each system must be tailored to satisfy site-specific conditions and operator requirements. Esso's continuing commitment to subsea technology development is currently exemplified by Esso UK's EDIPS project, which is developing building blocks that can be arranged in different ways to meet varying requirements.